Digital Watermarking and Steganography

Fundamentals and Techniques

Digital Watermarking and Steganography

FUNDAMENTALS AND TECHNIQUES

Frank Y. Shih

CRC Press
Taylor & Francis Group
Boca Raton London New York

CRC Press is an imprint of the
Taylor & Francis Group, an **Informa** business

CRC Press
Taylor & Francis Group
6000 Broken Sound Parkway NW, Suite 300
Boca Raton, FL 33487-2742

© 2008 by Taylor & Francis Group, LLC
CRC Press is an imprint of Taylor & Francis Group, an Informa business

No claim to original U.S. Government works
Printed in the United States of America on acid-free paper
10 9 8 7 6 5 4 3 2 1

International Standard Book Number-13: 978-1-4200-4757-8 (Hardcover)

Library of Congress Cataloging-in-Publication Data

Shih, Frank Y.
  Digital watermarking and steganography : fundamentals and techniques / Frank Y. Shih.
      p. cm.
  Includes bibliographical references and index.
  ISBN-13: 978-1-4200-4757-8 (alk. paper)
  ISBN-10: 1-4200-4757-4 (alk. paper)
   1. Computer security. 2. Multimedia systems--Security measures. 3. Intellectual property. I. Title.

QA76.9.A25S467 2007
005.8--dc22   2007034224

Visit the Taylor & Francis Web site at
http://www.taylorandfrancis.com

and the CRC Press Web site at
http://www.crcpress.com

Dedication

---

*To my loving wife and children, to my parents
who encouraged me through the years,
and to those who helped me in the process
of writing this book*

Table of Contents

Preface

Digital watermarking and steganography is an important topic because digital multimedia is widely used and the Internet is rapidly growing. This book intends to provide a comprehensive overview on the different aspects of mechanisms and techniques for information security. It is written for students, researchers, and professionals who take the related courses, want to improve their knowledge, and want to learn experiences pertaining to the role of digital watermarking and steganography.

Digital watermarking technology can be used to guarantee authenticity and can be applied as proof that the content has not been altered since insertion. Steganographic messages are often first encrypted by some traditional means, and then a covert text is modified in some way to contain the encrypted message. The need of information security exists everywhere everyday.

This book aims to provide students, researchers, and professionals with the technical information regarding digital watermarking and steganography, as well as instruct them in the fundamental theoretical framework in developing the extensive advanced techniques. By comprehensively considering the essential principles of the digital watermarking and steganographic systems, one cannot only obtain novel ideas in implementing the advanced algorithms, but also discover the new problems. The principles of digital watermarking and steganography in this book are illustrated with plentiful graphs and examples in order to simplify the problems, so readers can easily understand even complicated theories.

Several robust algorithms that are presented in this book to illustrate the framework provide assistance and tools in understanding and implementing the fundamental principles. The combinational spatial and frequency domains watermarking technique provides a new concept of enlarging the embedding capacity of watermarks. The genetic algorithm (GA) based watermarking technique solves the rounding error problem and provides an efficient embedding approach. The adjusted-purposed watermarking technique simplifies the selection and can be integrated into other watermarking techniques. The robust high-capacity watermarking technique successfully enlarges the hiding capacity while maintaining the watermark robustness. The GA-based steganography provides a new way of developing a robust steganographic system by artificially counterfeiting statistic features instead of the traditional strategy by avoiding the change of statistic features.

OVERVIEW OF THE BOOK

In chapter 1 digital watermarking and digital steganography are briefly introduced. Then the difference between watermarking and steganography is addressed. Next, a brief history along with updates of the resources of recent book publications is provided. The rest of the book is broken into two parts. Chapters 2–9 cover digital watermarking, and chapters 10–12 cover digital steganography.

In chapter 2 the classification of digital watermarking techniques based on characteristics is categorized into five types: blind versus nonblind, perceptible versus imperceptible, private versus public, robust versus fragile, and spatial domain versus frequency domain. The classification of digital watermarking techniques based on applications consists of five types: copyright protection watermarks, data authentication watermarks, fingerprint watermarks, copy control watermarks, and device control watermarks. In chapter 3 the basic mathematical preliminaries are introduced, including least significant bit substitution, discrete Fourier transform, discrete cosine transform, discrete wavelet transform, random sequence generation, chaotic map, error correction code, and set partitioning in the hierarchical tree. The fundamentals of digital watermarking are introduced in chapter 4. It is divided into four classes: spatial domain, frequency domain, fragile watermark, and robust watermark. In the spatial domain watermarking class, substitution watermarking and additive watermarking are introduced. In the frequency domain watermarking class, substitution watermarking, multiplicative watermarking, vector quantization watermarking, and the rounding error problem are introduced. In the fragile watermark class, the block-based fragile watermark, the weakness of the block-based fragile watermark, and the hierarchical-based fragile watermark are described. In the robust watermark, the redundant embedding approach and spread spectrum are discussed.

The issue of watermarking attacks is explored in chapter 5. The attacks are summarized into four types: image processing attacks, geometric attack, cryptographic attack, and protocol attack. Then the image processing attacks are divided into four classes: attack by filtering, attack by remodulation, attack by JPEG coding distortion, and attack by JPEG 2000 compression. Next, the geometric attacks are divided into nine classes: attack by image scaling, attack by rotation, attack by image clipping, attack by linear transformation, attack by bending, attack by warping, attack by perspective projection, attack by collage, and attack by templates. After that, the cryptographic attack and the protocol attack are explained. In the end, the available watermarking tools are provided. In chapter 6 the technique of combinational domain digital watermarking is introduced. The combination of spatial domain and frequency domain is described, and its advantages and experimental results are provided. The further encryption of combinational watermarking are explained. Chapter 7 shows how the genetic algorithm (GA) can be applied to digital watermarking. The concept and basic operations of the genetic algorithm are introduced and the fitness functions are discussed. Then the GA-based rounding error correction watermarking is introduced. Next, an application of the GA-based algorithm to medical image watermarking is presented.

The technique of adjusted-purpose digital watermarking is introduced in chapter 8. Following the overview, the morphological approach for extracting pixel-based features, the strategies for adjusting the varying sized transform window (VSTW), and the quantity factor (QF) are presented. The strategies of how VSTW is used to determine whether the embedded strategy should be in spatial or frequency domains and how QF is used to choose fragile, semifragile or robust watermarks are explained. Chapter 9 introduces a technique for robust high capacity digital watermarking. After the weaknesses of current robust watermarking are pointed out, the concept of robust watermarking is introduced. Following this, the processes of enlarging significant coefficients and breaking the local spatial similarity are explained. Next, the new concept of block-based chaotic map and the determination of embedding locations are presented. The design of the intersection-based pixels collection, the reference register, and the container is described. The robust high capacity watermarking algorithm and its embedding and extracting procedures are introduced. Finally, experimental results are provided to explore capacity enlargement, robust experiments, and performance comparisons.

Chapters 10–12 cover the topic of digital steganography. In chapter 10 three types of steganography are introduced: technical steganography, linguistic steganography, and digital steganography. Some applications of steganography are illustrated, including convert communication and one-time pad communication. The concern of embedding security and imperceptibility is explained. Four examples of steganography software are given: S-Tools, StegoDos, EzStego, and JSteg-Jpeg. Chapter 11 introduces the idea of steganalysis which intends to attack against steganography. The image statistical properties, the visual steganalytic system, and the IQM-based steganalytic system are described. Three learning strategies, the support vector machines, the neural networks, and the principle component analysis, are reviewed. Next, the frequency-domain steganalytic system is presented. The technique of genetic algorithm based steganography for breaking steganalytic systems is introduced in chapter 12. The emphasis is shifted from traditionally avoiding the change of statistic features to artificially counterfeiting the statistic features. An overview of the GA-based breaking methodology is first presented. Then the GA-based breaking algorithm on the spatial-domain steganalytic system (SDSS) is described. How one can generate the stego-image on the visual steganalytic system (VSS) and how one can generate the stego-image on the IQM-based steganalytic system are explained. Next, the strategy of the GA-based breaking algorithm on the frequency-domain steganalytic system (FDSS) is provided. Experimental results are shown that this algorithm cannot only pass the detection of steganalytic systems, but also increase the capacity of the embedded message and enhance the peak signal-to-noise ratio of stego-images.

FEATURES OF THE BOOK

- new state-of-the-art techniques for digital watermarking and steganography
- numerous practical examples
- a more intuitive development and a clear tutorial to the complex technology
- updated bibliography
- extensive discussion of watermarking and steganography
- inclusion of staganalysis techniques and their counter-examples.

FEEDBACK ON THE BOOK

It is my hope that an opportunity is given to correct any errors in this book; therefore, please provide a clear description of any errors that you may find. Your suggestions are always welcome on how to improve the textbook. For this, please use either email (shih@njit.edu) or regular mail to the author: Frank Y. Shih, College of Computing Sciences, New Jersey Institute of Technology, University Heights, Newark, NJ 07102-1982.

Acknowledgments

Portions of this book appeared in earlier forms as conference papers, journal papers, or theses with my students here at the New Jersey Institute of Technology. Therefore, these parts of the text are sometimes a combination of my words and those of my students. In particular, I drew some chapters from the work with my advised Ph.D. student, Yi-Ta Wu. I am indebted to him for writing the code and producing figures and images.

I would like to gratefully acknowledge the following publishers for giving me permission to re-use texts and figures that appeared in some of my past publications: the Institute of Electrical and Electronic Engineers (IEEE) and Elsevier Publishers.

Frank Y. Shih
New Jersey Institute of Technology

The Author

Frank Y. Shih earned a B.S. degree from National Cheng-Kung University, Taiwan, in 1980, an M.S. from the State University of New York at Stony Brook, in 1984, and a Ph.D. from Purdue University, West Lafayette, Indiana, in 1987, all in electrical and computer engineering. He is presently a professor jointly appointed in the Department of Computer Science, the Department of Electrical and Computer Engineering, and the Department of Biomedical Engineering at New Jersey Institute of Technology, Newark, New Jersey. He also serves as the director of the Computer Vision Laboratory.

Dr. Shih is currently on the editorial boards of the *International Journal of Pattern Recognition*, the *International Journal of Pattern Recognition Letters*, the *International Journal of Pattern Recognition and Artificial Intelligence*, the *International Journal of Recent Patents on Engineering*, the *International Journal of Recent Patents on Computer Science*, the *International Journal of Internet Protocol Technology*, and the *Journal of Internet Technology*. He has contributed as a steering member, committee member, and session chair for numerous professional conferences and workshops. Dr. Shih was the recipient of the Research Initiation Award from the National Science Foundation in 1991. He won the Honorable Mention Award from the International Pattern Recognition Society for Outstanding Paper and also won the Best Paper Award in the International Symposium on Multimedia Information Processing. He has received several awards for distinguished research at the New Jersey Institute of Technology and has served several times on the Proposal Review Panel of the National Science Foundation.

Prof. Shih started the mathematical morphology research with applications to image processing, feature extraction, and object representation. His *IEEE Transactions on Pattern Analysis and Machine Intelligence (PAMI)* article "Threshold Decomposition of Grayscale Morphology into Binary Morphology" is a breakthrough to solving the bottleneck problem in grayscale morphological processing. His several *IEEE Image Processing* and *IEEE Signal Processing* articles are innovations to achieve fast exact Euclidean distance transformation and to conduct robust image enhancement and segmentation using his developed "recursive soft morphological operators."

Prof. Shih further advanced the field of solar image processing and feature detection. Cooperating with physics researchers, he has made enormous contributions to fill in the gaps between solar physics and computer science. They have used the innovative computation and information technologies for real-time space weather monitoring and forecasting, and have received a National Science Foundation grant of more than $1 million. They have developed several methods to automatically detect and characterize filament/prominence eruptions, flares, and coronal mass ejections (CMEs). These techniques are currently in use at the Big Bear Observatory in California as well as by NASA.

Dr. Shih has also made significant contributions to information hiding, focusing on the security and robustness of digital steganography and watermarking. He has developed several novel methods to increase embedding capacity, enhance robustness, integrate different watermarking platforms, and break the steganalytic systems. His recent article, published in *IEEE Transactions on Systems, Man, and Cybernetics (SMC)*, is the first one to apply genetic algorithm-based methodology to break the steganalytic systems.

Dr. Shih has published 90 journal papers, 90 conference papers, and 7 book chapters. He has overcome many difficult research problems in multimedia signal processing, pattern recognition, features extraction, and information security. Some examples are the robust information hiding, automatic solar features classification, optimum features reduction, fast accurate Euclidean distance transform, and fully parallel thinning algorithm.

Dr. Shih holds the research fellow for the American Biographical Institute and the IEEE senior membership. His current research interests include digital watermarking and steganography, digital forensics, image processing, computer vision, sensor networks, pattern recognition, bioinformatics, information security, robotics, fuzzy logic, and neural networks.

1 Introduction

Digital information and data are transmitted more often over the Internet now than ever before. The availability and efficiency of global computer networks for the communication of digital information and data have accelerated the popularity of digital media. Digital images, video, and audio have been revolutionized in the way they can be captured, stored, transmitted, and manipulated, and this gives rise to a wide range of applications in education, entertainment, the media, medicine, and the military, among other fields [1].

Computers and networking facilities are becoming less expensive and more widespread. Creative approaches to storing, accessing, and distributing data have generated many benefits for the digital multimedia field, mainly due to properties such as distortion-free transmission, compact storage, and easy editing. But free-access digital multimedia communication unfortunately also provides virtually unprecedented opportunities to pirate copyrighted material. Therefore, the idea of using a digital watermark to detect and trace copyright violations has stimulated significant interests among engineers, scientists, lawyers, artists, and publishers, to name a few. As a result, the research in watermark embedding robustness with respect to compression, image-processing operations, and cryptographic attacks has become very active in recent years, and the developed techniques have grown and been improved a great deal.

1.1 DIGITAL WATERMARKING

Watermarking is not a new phenomenon. For nearly a thousand years, watermarks on paper have been used to visibly indicate a particular publisher and to discourage counterfeiting in currency. A watermark is a design impressed on a piece of paper during production and used for copyright identification (as illustrated in Figure 1.1). The design may be a pattern, a logo, or some other image. In the modern era, as most data and information are stored and communicated in digital form, proving authenticity plays an increasingly important role. As a result, digital watermarking is a process whereby arbitrary information is encoded into an image in such a way as to be is imperceptible to image observers.

Digital watermarking has been proposed as a suitable tool for identifying the source, creator, owner, distributor, or authorized consumer of a document or an image. It can also be used to detect a document or an image that has been illegally distributed or modified. Another technology, encryption, is a process of obscuring information to make it unreadable to observers without specific keys or knowledge. This technology is sometimes referred to as *data scrambling*. Watermarking, when complemented by encryption, can serve a vast number of purposes including copyright protection, broadcast monitoring, and data authentication.

FIGURE 1.1 A paper watermark.

In the digital world, a watermark is a pattern of bits inserted into a digital media that can identify the creator or authorized users. The digital watermark—unlike the traditional printed, visible watermark—is designed to be invisible to viewers. The bits embedded into an image are scattered all around to avoid identification or modification. Therefore, a digital watermark must be robust enough to survive the detection, compression, and other operations that might be applied upon a document.

Figure 1.2 depicts a general digital watermarking system. A watermark message W is embedded into a media message, which is defined as the host image H. The resulting image is the watermarked image H^*. In the embedding process, a secret key K—that is, a random number generator—is sometimes involved to generate a more secure watermark. The watermarked image H^* is then transmitted along a communication channel. The watermark can later be detected or extracted by the recipient.

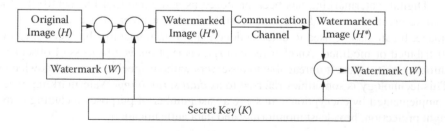

FIGURE 1.2 A general digital watermarking system.

Imperceptibility, security, capacity, and robustness are among the many aspects of watermark design. The watermarked image must look indistinguishable from the original image; if a watermarking system distorts the host image to some point of being perceptible, it is of no use. An ideal watermarking system should embed a large amount of information perfectly securely, but with no visible degradation to the host image. The embedded watermark should be robust with invariance to intentional (e.g., noise) or unintentional (e.g., image enhancement, cropping, resizing, or compression) attacks. Many researchers have been focusing on security and robustness, but rarely on watermarking capacity [2–3]. The amount of data an algorithm can embed in an image has implications for how the watermark can be applied. Indeed, both security and robustness are important because the embedded watermark is expected to be imperceptible and unremovable. Nevertheless, if a large watermark can be embedded into a host image, the process could be useful for many other applications.

Another scheme is the use of keys to generate random sequences during the embedding process. In this scheme, the cover image (i.e., the host image) is not needed during the watermark detection process. It is also a goal that the watermarking system utilizes an asymmetric key, as in public or private key cryptographic systems. (A public key is used for image verification and a private key is needed for embedding security features.) Knowledge of the public key neither helps compute the private key nor allows removal of the watermark.

According to user embedding purposes, watermarks can be categorized into three types: *robust*, *semifragile*, and *fragile*. Robust watermarks are designed to withstand arbitrary, malicious attacks such as image scaling, bending, cropping, and lossy compression [4–7]. They are usually used for copyright protection in order to declare rightful ownership. Semifragile watermarks are designed for detecting any unauthorized modification, while at the same time allowing some image-processing operations [8]. In other words, it is selective authentication that detects illegitimate distortion while ignoring applications of legitimate distortion. For the purpose of image authentication, fragile watermarks [9–13] are adopted to detect any unauthorized modification at all.

In general, we can embed a watermark in two types of domains: the spatial domain or the frequency domain [14–17]. In the spatial domain we can replace the pixels in the host image with the pixels in the watermarked image [7, 18]. Note that a sophisticated computer program may easily detect the inserted watermark. In the frequency domain we can replace the coefficients of a transformed image with the pixels in the watermarked image [19, 20]. (The frequency-domain transformations most commonly used are discrete cosine transform, discrete Fourier transform, and discrete wavelet transform.) This kind of embedded watermark is, in general, difficult to detect. However, its embedding capacity is usually low, since a large amount of data will distort the host image significantly. The watermark must be smaller than the host image; in general, the size of a watermark is one-sixteenth the size of the host image.

1.2   DIGITAL STEGANOGRAPHY

Digital steganography aims at hiding digital information into covert channels so that one can conceal the information and prevent the detection of the hidden message. Steganalysis is the art of discovering the existence of hidden information; as such,

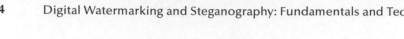

FIGURE 1.3 A classic steganographic model.

steganalytic systems are used to detect whether an image contains a hidden message. By analyzing various image features between stego-images (images containing hidden messages) and cover images (images containing no hidden messages), a steganalytic system is able to detect stego-images. Cryptography is the practice of scrambling a message to an obscured form to prevent others from understanding it, while steganography is the practice of obscuring the message so that it cannot be discovered.

Figure 1.3 depicts a classic steganographic model presented by Simmons [21]. In it, Alice and Bob are planning to escape from jail. All communications between them are monitored by the warden Wendy, so they must hide their messages in other innocuous-looking media (cover objects) in order to obtain each others' stego-images. The stego-images are then sent through public channels. Wendy is free to inspect all messages between Alice and Bob in one of two ways: passively or actively. The passive approach involves inspecting the message in order to determine whether it contains a hidden message and then to conduct a proper action. The active approach involves always altering Bob's and Alice's messages even if Wendy may not perceive any traces of hidden meaning. Examples of the active method would be image-processing operations such as lossy compression, quality-factor alteration, format conversion, palette modification, and low-pass filtering.

For steganographic systems, the fundamental requirement is that the stego-image be perceptually indistinguishable to the degree that it does not raise suspicion. In other words, the hidden information introduces only slight modification to the cover object. Most passive wardens detect the stego-images by analyzing their statistic features. In general, steganalytic systems can be categorized into two classes: spatial-domain steganalytic systems (SDSSs) and frequency-domain steganalytic systems (FDSSs). The SDSS [22, 23] is adopted for checking lossless compressed images by analyzing the spatial-domain statistic features. For lossy compressed images, such as a JPEG file, the FDSS is used to analyze frequency-domain statistic features [24, 25]. Westfeld and Pfitzmann have presented two SDSSs based on visual and chi-square attacks [23]. The visual attack uses human eyes to inspect stego-images by checking their lower bit planes, while the chi-square attack can automatically detect the specific characteristic generated by the least-significant-bit steganographic technique.

1.3 DIFFERENCES BETWEEN WATERMARKING AND STEGANOGRAPHY

Watermarking is closely related to steganography; however, there are some differences between the two. Watermarking mainly deals with image authentication, whereas steganography deals with hiding data. Embedded watermarking messages usually pertain to host image information, such as copyright, so they are bound with the cover image. Watermarking is often used whenever the cover image is available to users who are aware of the existence of the hidden information and may intend to remove it. Hidden messages in steganography are usually not related to the host image. They are designed to make extremely important information imperceptible to any interceptors.

In watermarking, the embedded information is related to an attribute of the carrier and conveys additional information or properties about the carrier. The primary object of the communication channel is the carrier itself. In steganography, the embedded message usually has nothing to do with the carrier, which is simply used as a mechanism to pass the message. The object of the communication channel is the hidden message. As the application of watermarking, a balance between image perceptual quality and robustness is maintained. Constraints in maintaining image quality tend to reduce the capacity of information embedded. As the application of steganography is different, dealing covert message transfer, the embedded capacity is often viewed with as much importance as robustness and image quality.

1.4 A BRIEF HISTORY

The term *watermarking* is derived from the history of traditional papermaking. Wet fiber is pressed to expel the water, and the enhanced contrast between watermarked and nonwatermarked areas of the paper forms a particular pattern and becomes visible.

Watermarking originated in the paper industry in the late Middle Ages—roughly the 13th century. The earliest known usage appears to record the paper brand and the mill that produced it so that authenticity could be clearly recognized. Later, watermarking was used to certify the composition of paper. Nowadays, many countries watermark their paper, currencies, and postage stamps to make counterfeiting more difficult.

The digitization of our world has supplemented traditional watermarking with digital forms. While paper watermarks were originally used to differentiate between different paper manufacturers, today's digital watermarks have more widespread uses. Stemming from the legal need to protect the intellectual property of the creator from unauthorized usage, digital watermarking technology attempts to reinforce the copyright by embedding a digital message that can identify the creator or the intended recipients. When encryption is broken, watermarking is essentially the technology to protect unencrypted multimedia content.

In 1989, Komatsu and Tominaga proposed digital watermarking to detect illegal copies [26]. They encoded a secret label into a copy using slight modification on redundant information. When the label matches that of the registered owner, the provider can ensure that the document holder is the same person. As a method, digital watermarking has a long history, but it was only after 1990 that it gained large international interest. Today, a great number of conferences and workshops on this topic are held, and there are a large number of scientific journals on watermarking in publication. This renewed scientific interest in digital watermarking has very quickly grabbed the attention of industry. The widely used application of it includes copyright protection, labeling, monitoring, tamper proofing, and conditional access.

Watermarking or information embedding is a particular embodiment of steganography. The term *steganography* is derived from the Greek word for "covered or hidden writing." It is intended to hide the information in a medium in such a manner that no one except the anticipated recipient knows the existence of the information. This is in contrast to cryptography, which focuses on making information unreadable to any unauthorized persons.

The history of steganography can be traced back to ancient Greece, where the message-hiding procedure included tattooing a shaved messenger's head, waiting for his hair to grow back, and then sending him out to deliver the message personally; the recipient would then shave the messenger's head once again in order to read the message. Another procedure included etching messages in wooden tablets and covering them with wax. Various types of steganography and cryptography also thrived in ancient India, and in ancient China, military generals and diplomats hid secret messages on thin sheets of silk or paper. One famous story on the successful revolt of the Han Chinese against the Mongolians during the Yuan dynasty demonstrates a steganographic technique: During the Yuan dynasty (A.D. 1280–1368), China was ruled by the Mongolians. On the occasion of the Mid-Autumn Festival, the Han people made moon cakes (as cover objects), with a message of an attack plan inside (as hidden information). The moon cakes were distributed to members to inform them of the planned revolt, which successfully overthrew the Mongolian regime.

REFERENCES

[1] Berghel, H., and L. O'Gorman. "Protecting Ownership Rights through Digital Watermarking." *IEEE Computer Mag.* 29 (1996): 101.

[2] Barni, M., F. Bartolini, A. De Rosa, and A. Piva. "Capacity of the Watermark Channel: How Many Bits Can Be Hidden within a Digital Image?" In *Security and Watermarking of Multimedia Contents*, ed. P. W. Wong and E. J. Delp III. Proc. SPIE 2657, San Jose, CA: SPIE, 1999.

[3] Shih, F. Y., and S. Y. Wu. "Combinational Image Watermarking in the Spatial and Frequency Domains." *Pattern Recognition* 36 (2003): 969.

[4] Cox, I., et al.. "Secure Spread Spectrum Watermarking for Images Audio and Video." In *Proc. IEEE Int. Conf. Image Processing*. Lausanne, Switzerland: IEEE, 1996.

[5] Cox, I., et al. "Secure Spread Spectrum Watermarking for Multimedia." *IEEE Trans. Image Processing* 6 (1997): 1673.

[6] Lin, S. D., and C.-F. Chen. "A Robust DCT-Based Watermarking for Copyright Protection." *IEEE Trans. Consumer Electronics* 46 (2000): 415.

[7] Nikolaidis, N., and I. Pitas. "Robust Image Watermarking in the Spatial Domain." *Signal Processing* 66 (1998): 385.

[8] Acharya, U. R., et al. "Compact Storage of Medical Image with Patient Information." *IEEE Trans. Info. Tech. in Biomedicine* 5 (2001): 320.

[9] Caronni, G. "Assuring Ownership Rights for Digital Images." In *Proc. Reliable IT Systems*, Germany: Viewveg, 1995.

[10] Celik, M. et al. "Hierarchical Watermarking for Secure Image Authentication with Localization." *IEEE Trans. Image Processing* 11 (2002): 585.

[11] Pitas, I., and T. Kaskalis. "Applying Signatures on Digital Images." In *Proc. IEEE Workshop Nonlinear Signal and Image Processing*, 1995.

[12] Wolfgang, R., and E. Delp. "A Watermarking Technique for Digital Imagery: Further Studies." In *Proc. Int. Conf. Imaging Science, Systems and Technology*, Las Vegas, NV, 1997.

[13] Wong, P. W. "A Public Key Watermark for Image Verification and Authentication." In *Proc. IEEE Int. Conf. Image Processing*, Chicago, 1998.

[14] Langelaar, G., et al. "Watermarking Digital Image and Video Data: A State-of-the-Art Overview." *IEEE Signal Processing* 17 (2000): 20.

[15] Cox, I. J., et al. "Secure Spread Spectrum Watermarking for Multimedia." *IEEE Trans. Image Processing* 6 (1997): 1673.

[16] Petitcolas, F., R. Anderson, and M. Kuhn. "Information Hiding—A Survey. *Proc. of the IEEE* 87 (1999): 1062.

[17] Cox, I., and M. Miller. "The First 50 Years of Electronic Watermarking." *J. Appl. Signal Processing* 2 (2002): 126.

[18] Bruyndonckx, O., J.-J. Quisquater, and B. Macq. "Spatial Method for Copyright Labeling of Digital Images." In *Proc. IEEE Workshop Nonlinear Signal and Image Processing*, Neos Marmaras, Greece, 1995.

[19] Huang, J., Y. Q. Shi, and Y. Shi. "Embedding Image Watermarks in DC Components." *IEEE Trans. Circuits and Systems for Video Technology* 10 (2000): 974.

[20] Lin, S. D., and C.-F. Chen. "A Robust DCT-Based Watermarking for Copyright Protection." *IEEE Trans. Consumer Electronics* 46 (2000): 415.

[21] Simmons, G. J. "Prisoners' Problem and the Subliminal Channel." In *Proc. Int. Conf. Advances in Cryptology*, 1984.

[22] Avcibas, I., N. Memon, and B. Sankur. "Steganalysis Using Image Quality Metrics." *IEEE Trans. Image Processing* 12 (2003): 221.

[23] Westfeld, A., and A. Pfitzmann. "Attacks on Steganographic Systems Breaking the Steganographic Utilities EzStego, Jsteg, Steganos, and S-Tools and Some Lessons Learned." In *Proc. Int. Workshop Information Hiding*, Dresden, Germany, 1999, 61.

[24] Farid, H. *Detecting Steganographic Message in Digital Images*. Technical Report TR2001-412, Computer Science. Hanover, NH: Dartmouth College, 2001.

[25] Fridrich, J., M. Goljan, and D. Hogea. "New Methodology for Breaking Steganographic Techniques for JPEGs." In *Proc. SPIE*, Santa Clara, CA: SPIE, 2003.

[26] Komatsu, N., and H. Tominaga. "A Proposal on Digital Watermark in Document Image Communication and Its Application to Realizing a Signature." *Trans. of the Institute of Electronics, Information and Communication Engineers* J72B-I (1989): 208.

APPENDIX: SELECTED LIST OF BOOKS ON WATERMARKING AND STEGANOGRAPHY

Arnold, M., S. Wolthusen, and M. Schmucker. *Techniques and Applications of Digital Watermarking and Content Protection*. Norwood, MA: Artech House, 2003.

Baldoza, A. *Data Embedding for Covert Communications, Digital Watermarking, and Information Augmentation*. Storming Media, 2000.

Barni, M., and F. Bartolini. *Watermarking Systems Engineering: Enabling Digital Assets Security and Other Applications.* Boca Raton, FL: CRC Press, 2004.

Chandramouli, R. *Digital Data-Hiding and Watermarking with Applications.* Boca Raton, FL: CRC, 2003.

Cox, I., M. Miller, and J. Bloom. *Digital Watermarking: Principles and Practice.* San Francisco: Morgan Kaufmann, 2001.

Eggers, J., and B. Girod. *Informed Watermarking.* Springer, 2002.

Furht, B., and D. Kirovski. *Multimedia Watermarking Techniques and Applications.* Boca Raton, FL: Auerbach, 2006.

Furht, B., E. Muharemagic, and D. Socek. *Multimedia Encryption and Watermarking.* New York: Springer, 2005.

Gaurav, R. *Digital Encryption Model Using Embedded Watermarking Technique in ASIC Design.* Ann Arbor, MI: ProQuest/UMI, 2006.

Johnson, N., Z. Duric, and S. Jajodia. *Information Hiding: Steganography and Watermarking—Attacks and Countermeasures.* Norwell, MA: Kluwer Academic, 2001.

Katzenbeisser, S., and F. Petitcolas. *Information Hiding Techniques for Steganography and Digital Watermarking.* Norwood, MA: Artech House, 2000.

Kipper, G. *Investigator's Guide to Steganography.* Boca Raton, FL: CRC, 2003.

Kwok, S., C. Yang, K. Tam, and J. Wong. *Intelligent Watermarking Techniques.* World Scientific, 2004.

Lu, C. *Multimedia Security: Steganography and Digital Watermarking Techniques for Protection of Intellectual Property.* Idea Group, 2005.

Pan, J., H. Huang, and L. Jain. *Intelligent Watermarking Techniques.* World Scientific, 2004.

Pfitzmann, A. *Information Hiding.* Springer, 2000.

Seitz, J. *Digital Watermarking for Digital Media.* Idea Group Inc, 2005.

Su, J., *Digital Watermarking Explained.* New York: Wiley, 2003.

Wayner, P. *Disappearing Cryptography,* 2d ed.: *Information Hiding: Steganography and Watermarking.* San Francisco: Morgan Kaufmann, 2002.

2 Classification in Digital Watermarking

With the rapidly increasing number of electronic commerce web sites and applications, intellectual property protection is an extremely important concern for content owners who exhibit digital representations of photographs, books, manuscripts, and original artwork on the Internet. Moreover, as available computing power continues to rise, there is an increasing interest in protecting video files from alteration. Digital watermarking's application is widely spread across electronic publishing, advertisng, merchandise ordering and delivery, picture galleries, digital libraries, online newspapers and magazines, digital video and audio, personal communication, and more.

Digital watermarking is one of the technologies being developed as a suitable tool for identifying the source, creator, owner, distributor, or authorized consumer of a document or an image. It can also be used for tracing images that have been illegally distributed. There are two major steps in the digital watermarking process: (1) *watermark embedding*, in which a watermark is inserted into a host image; and (2) *watermark extraction*, in which the watermark is pulled out of the image. Because there are a great number of watermarking algorithms developed, it is important to define some criteria for classifying them in order to understand how different watermarking schemes can be applied.

2.1 CLASSIFICATION BASED ON CHARACTERISTICS

This section categorizes digital watermarking technologies into five classes according to the characteristics of embedded watermarks:

1. blind versus nonblind
2. perceptible versus imperceptible
3. private versus public
4. robust versus fragile
5. spatial domain-based versus frequency domain-based

2.1.1 BLIND VERSUS NONBLIND

A watermarking technique is said to be *blind* if it does not require access to the original unwatermarked data (image, video, audio, etc.) to recover the watermark. Conversely, a watermarking technique is said to be *nonblind* if the original data is needed for the extraction of the watermark. For example, the algorithms of Barni et al. and Nikolaidis and Pitas belong to the blind category, while those of Cox et al. and Swanson et al. belong to the nonblind [1–4]. In general, the nonblind scheme is more robust than the blind one because it is obvious that the watermark can be extracted easily by knowing the unwatermarked data. However, in most applications, the unmodified host signal is not available to the watermark detector. Since the blind scheme does not need the original data, it is more useful than the nonblind one in most applications.

2.1.2 PERCEPTIBLE VERSUS IMPERCEPTIBLE

A watermark is said to be *perceptible* if the embedded watermark is intended to be visible—for example, a logo inserted into a corner of an image. A good perceptible watermark must be difficult for an unauthorized person to remove and can resist falsification. Since it is relatively easy to embed a pattern or a logo into a host image, we must make sure the perceptible watermark was indeed the one inserted by the author. In contrast, an *imperceptible* watermark is embedded into a host image by sophisticated algorithms and is invisible to the naked eye. It could, however, be extracted by a computer.

2.1.3 PRIVATE VERSUS PUBLIC

A watermark is said to be *private* if only authorized users can detect it. In other words, private watermarking techniques invest all efforts to make it impossible for unauthorized users to extract the watermark, for instance, using a private, pseudorandom key. This private key indicates a watermark's location in the host image, allowing insertion and removal of the watermark if the secret location is known. In contrast, watermarking techniques that allow anyone to read the watermark are called *public*. Public watermarks are embedded in a location known to everyone, so the watermark detection software can easily extract the watermark by scanning the whole image. In general, private watermarking techniques are more robust than public ones, in which an attacker can easily remove or destroy the message once the embedded code is known.

There is also the *asymmetric* form of public watermarking, through which any user can read the watermark without being able to remove it. This is referred to as an *asymmetric cryptosystem*. In this case, the detection process (and in particular the detection key) is fully known to anyone, so only a public key is needed for verification and a private key is used for the embedding.

2.1.4 ROBUST VERSUS FRAGILE

Watermark robustness accounts for the capability of the hidden watermark to survive legitimate daily usage or image-processing manipulation, such as intentional or unintentional attacks. Intentional attacks aim at destroying the watermark, while unintentional attacks do not explicitly intend to alter it. According to embedding purposes, watermarks can be categorized into three types: (1) robust, (2) semifragile, and (3) fragile. *Robust* watermarks are designed to survive intentional (malicious) and unintentional (nonmalicious) modifications of the watermarked image [2, 3, 5, 6]. Unfriendly intentional modifications include unauthorized removal or alternation of the embedded watermark and unauthorized embedding of any other information. Unintentional modifications include image processing operations such as scaling, cropping, filtering, and compression. Robust watermarks are usually used for copyright protection, to declare rightful ownership.

Semifragile watermarks are designed for detecting any unauthorized modification, at the same time allowing some image-processing operations [7]. They are used for selective authentication that detects illegitimate distortion while ignoring applications of legitimate distortion. In other words, semifragile watermarking techniques can discriminate common image processing and small-content-preserving noise—such as lossy compression, bit error, or "salt-and-pepper" noise—from malicious content modification.

For the purpose of authentication, *fragile* watermarks are adopted to detect any unauthorized modification [8–12]. fragile watermarking techniques are concerned with complete integrity verification. The slightest modification of the watermarked image will alter or destroy the fragile watermark.

2.1.5 SPATIAL DOMAIN-BASED VERSUS FREQUENCY DOMAIN-BASED

There are two image domains for embedding watermarks: the spatial domain and the frequency domain. In the spatial domain [13], we can simply insert a watermark into a host image by changing the gray value of some pixels in the host image. This has the advantages of low complexity and easy implementation, but the inserted information may be easily detected using computer analysis or could be easily attacked. We can embed the watermark into the coefficients of a transformed image in the frequency domain [3, 14]. The transformations include discrete cosine transform, discrete Fourier transform, and discrete wavelet transform. However, if we embed too much data in the frequency domain, the image quality will be degraded significantly.

Spatial domain watermarking techniques are usually less robust to attacks such as compression and noise. However, they have much lower computational complexity and usually can survive the cropping attack, which often the frequency domain watermarking techniques fail. Another technique is combining both spatial domain watermarking and frequency domain watermarking for increased robustness and less complexity.

2.2 CLASSIFICATION BASED ON APPLICATIONS

Digital watermarking techniques embed hidden information directly into the media data. In addition to cryptographic schemes, watermarking represents an efficient technology to ensure data integrity as well as data origin authenticity. Copyright, authorized recipients, or integrity information can be embedded using a secret key into a host image as a transparent pattern. This section categorizes digital watermarking technologies into five classes according to their applications:

1. copyright protection watermarks
2. data authentication watermarks
3. fingerprinting watermarks
4. copy control watermarks
5. device control watermarks

2.2.1 Copyright Protection Watermarks

A watermark is invisibly inserted into an image that can be detected when the image is compared with the original. This watermark for copyright protection is designed to identify both the source of the image as well as its authorized users. Public key encryption, such as the RSA algorithm [15], does not completely prevent unauthorized copying because of the ease with which images could be reproduced from previously published documents. All encrypted documents and images must be decrypted before the inspection. After the encryption is taken off, the document can be readable and disseminated. The idea of embedding an invisible watermark to identify ownership and recipients has attracted many interests in the printing and publishing industries.

Digital video can be copied repeatedly without loss of quality. Therefore, copyright protection for video data is more critical in digital video delivery networks than it was with analog TV broadcasting. One copyright protection method is to add a watermark to the video stream that carries information about the sender and recipient. In this way video watermarking can enable identification and tracing of different copies of video data. It can be applied to video distribution over the Internet, pay-per-view video broadcasting, and video disk labeling.

2.2.2 Data Authentication Watermarks

The traditional means of data authentication is applied to a document in the form of a handwritten signature. Since the signature is affixed to a document, it is difficult to be modified or transported to another document. The comparison of two handwritten signatures can determine whether they were created by the same person.

A digital signature replicates the handwritten signature and offers an even stronger degree of authentication. A user can sign a digital document by encrypting it with a private key and an encryption scheme. *Digital signatures* make use of a public key or asymmetric cryptography, in which two keys related to each other mathematically are used. One can verify the digital signatures using only the public key, but will need the secret key for the generation of digital signatures. Therefore, the public key is available to anyone who wishes to conduct verification, but the private key is merely given to authorized persons.

Data redundancy of an image makes it capable of authenticating the embedded watermark without accessing the original image. Although image format conversion leads to a different image representation, it does not change the visual appearance and the authenticity status. However, the authenticated image will inevitably be distorted by a small amount of noise due to the authentication itself. The distortion is often quite small, but this is usually unacceptable for medical imagery or images with a high strategic importance in military applications.

2.2.3 Fingerprint Watermarks

Biometrics technology, such as face, fingerprint, and iris recognition, plays an important role in today's personal identification systems. Digital watermarking of fingerprint images can be applied to protect the fingerprint images against malicious attacks, can detect fraud fingerprint images, and can ensure secure transmission. Fingerprinting in digital watermarking is usually used as the process of embedding the identity to an image in such a way that it is difficult to erase. This allows the copyright owner to trace pirates if the image is distributed illegally. In this usage, watermarking is a method of embedding hidden information, known as the *payload*, within content. The content could be audio, image, or video, while the payload could identify the content owner or usage permission for the content. The payload of fingerprint watermarking is an identification number unique to each recipient of the content, the aim being to determine the source of illegally distributed copies of the content.

2.2.4 COPY CONTROL WATERMARKS

IBM's Tokyo Research Laboratory (TRL) first proposed the use of watermarking technology for DVD copy protection at the DVD Copy Protection Technical Working Group (CPTWG) in September 1996, and showed that the TRL's technology can detect embedded watermarks even in the MPEG2-compressed domain. This development enables DVD playing and recording devices to automatically prevent playback of unauthorized copies and unauthorized recording using copy control information detected in digital video content. The digital data are transmitted from a transmitter-side apparatus to a receiver-side apparatus through interfaces that allow the transmission and receiving of the digital data only between the authenticated apparatuses. Copy control information indicating a copy restriction level is added to the main data recorded on the digital medium such that the main data contains a first portion containing image and/or voice information, and a second portion containing the copy control information. The digital watermark is embedded into the second portion of the main data.

2.2.5 DEVICE CONTROL WATERMARKS

Device control watermarks are embedded to control access to a resource using a verifying device. A watermarking system embeds an authorization code in a signal and transmits it to a verifying device (e.g., as a television or radio program). In the verifying device, the authorization code is extracted from the watermarked signal and an operation to be performed on the resource is authorized separately on the extracted authorization code, which may consist of permission for executing a program or copying a multimedia object.

These watermarks can also be embedded in an audio signal to remotely control a device such as a toy, a computer, or an appliance. The device is equipped with an appropriate detector to identify hidden signals, which can trigger an action or change a state of the device. These watermarks can be used with a time-gate device, where detection of the watermark holds a time interval within which a user is permitted to conduct an action such as typing or pushing a button.

REFERENCES

[1]   Barni, M., et al. "A DCT-Domain System for Robust Image Watermarking." *Signal Processing* 66 (1998): 357.

[2]   Nikolaidis, N., and I. Pitas. "Robust Image Watermarking in the Spatial Domain." *Signal Processing* 66 (1998): 385.

[3]   Cox, I. J., et al. "Secure Spread Spectrum Watermarking for Multimedia." *IEEE Trans. Image Processing*, 6 (1997): 1673.

[4]   Swanson, M. D., B. Zhu, and A. H. Tewfik. "Transparent Robust Image Watermarking." In *Proc. IEEE Int. Conf. on Image Processing*. Lausanne, Switzerland: IEEE, 1996.

[5]   Lin, S. D., and C.-F. Chen. "A Robust DCT-Based Watermarking for Copyright Protection." *IEEE Trans. Consumer Electronics* 46 (2000): 415.

[6]   Deguillaume, F., S. Voloshynovskiy, and T. Pun. "Secure Hybrid Robust Watermarking Resistant against Tampering and Copy Attack." *Signal Processing* 83 (2003): 2133.

[7]   Sun, Q., and S.-F. Chang. "Semifragile Image Authentication Using Generic Wavelet Domain Features and ECC." In *Proc. IEEE Int. Conf. on Image Processing*. 2002.

[8]   Wong, P. W. "A Public Key Watermark for Image Verification and Authentication." In *Proc. IEEE Int. Conf. Image Processing*. Chicago, 1998.

[9]   Celik, M. U., et al. "Hierarchical Watermarking for Secure Image Authentication with Localization." *IEEE Trans. Image Processing* 11 (2002): 585.

[10] Wolfgang, R., and E. Delp. "A Watermarking Technique for Digital Imagery: Further Studies." In *Proc. Int. Conf. Imaging Science, Systems and Technology*. Las Vegas, NV, 1997.

[11] Pitas, I., and T. Kaskalis. "Applying Signatures on Digital Images." In *Proc. IEEE Workshop on Nonlinear Signal and Image Processing*, Neos Marmaras, Greece, 1995, 460.

[12] Caronni, G. "Assuring Ownership Rights for Digital Images." In *Proc. Reliable IT Systems*. Germany: Viewveg, 1995.

[13] Berghel, H., and L. O'Gorman. "Protecting Ownership Rights through Digital Watermarking." *IEEE Computer Mag.* 29 (1996): 101.

[14] Cox, I., et al. "Secure Spread Spectrum Watermarking for Images Audio and Video." In *Proc. IEEE Int. Conf. Image Processing*. Lausanne, Switzerland: IEEE, 1996.

[15] Chambers, W. G. *"Basics of Communications and Coding."* Oxford Science Publications. Oxford: Clarendon Press, 1985.

3 Mathematical Preliminaries

This chapter introduces mathematical preliminaries for performing digital watermarking techniques for different embedding purposes and domains, and presents some commonly used operations in digital watermarking, including *least-significant-bit substitution, discrete Fourier transform, discrete cosine transform, discrete wavelet transform, random sequence generation, the chaotic map, error correction code,* and *set partitioning in hierarchical tree.*

3.1 LEAST-SIGNIFICANT-BIT SUBSTITUTION

Least-significant-bit (LSB) substitution is often used in image watermarking for embedding watermarks into a host image. In a grayscale image, a pixel is represented as 8 bits with the most significant bit (MSB) to the left and the least significant bit to the right. For example, a pixel having the gray value 130 is shown in Figure 3.1a. The idea of LSB substitution is to replace the LSB of a pixel with the watermark, because this has little effect on the appearance of the carrier message. For example, as shown in Figure 3.1b, when the LSB is changed, the pixel value changes from 130 to 131, which is undetectable in human perception. If we change bits other than the LSB, the image will be noticeably distorted. When we alter the bit closer to the MSB, the image will be distorted more. For instance, if we change the MSB, the pixel value 130 will be changed to 2, as shown in Figure 3.1c. This makes a significant change to the gray intensity view. Sometimes, in order to embed more watermarks, the least number of bits are replaced. Since only the lower-order bits are altered, the resulting color shifts are typically imperceptible [1].

In general, the image formats adopted in the LSB substitution are lossless. The LSB method is easy to implement and can possess high embedding capacity and low visual perceptibility because every cover bit contains one bit of the hidden message. However, because the data hidden in the LSB may be known, the LSB substitution method is impervious to watermark extraction and some attacks such as cropping and compression.

3.2 DISCRETE FOURIER TRANSFORM (DFT)

In the early 1800s, French mathematician Joseph Fourier introduced the Fourier series for the representation of the continuous-time periodic signal. The signal can be decomposed into a linear weighted sum of harmonically related complex exponentials. This weighted sum represents the frequency content of a signal, called the *spectrum.* When the signal becomes nonperiodic, its period becomes infinite and its spectrum becomes continuous. An image is considered as a spatially varying function. Fourier transform decomposes such an image function into a set of orthogonal functions, and can transform the spatial intensity image into its frequency domain. From continuous form, one can obtain the form for discrete time images.

(a) | 1 | 0 | 0 | 0 | 0 | 0 | 1 | 0 |

(b) | 1 | 0 | 0 | 0 | 0 | 0 | 1 | 1 |

(c) | 0 | 0 | 0 | 0 | 0 | 0 | 1 | 0 |

FIGURE 3.1 (a) An 8-bit pixel with a value of 130. (b) The value is changed to 131 after the LSB substitution. (c) The value is changed to 2 after the MSB substitution.

This section introduces the discrete case of two-dimensional (2D) Fourier transform. If $f(x,y)$ denotes a digital image in spatial domain and $F(u,v)$ denotes a transform image in frequency domain, the general equation for a 2D discrete Fourier transform (DFT) is defined as follows:

$$F(u,v) = \frac{1}{MN} \sum_{x=0}^{M-1} \sum_{y=0}^{N-1} f(x,y) \exp\left[-j2\pi\left(\frac{ux}{M} + \frac{vy}{N}\right)\right], \tag{3.1}$$

for $u = 0,1,2,\ldots,M-1$ and $v = 0,1,2,\ldots,N-1$. The inverse DFT can be represented as

$$f(x,y) = \sum_{u=0}^{M-1} \sum_{v=0}^{N-1} F(u,v) \exp\left[j2\pi\left(\frac{ux}{M} + \frac{vy}{N}\right)\right], \tag{3.2}$$

for $x = 0,1,2,\ldots,M-1$ and $y = 0,1,2,\ldots,N-1$. Because $F(u,v)$ and $f(x,y)$ are a Fourier transform pair, the grouping of the constant terms is not important. In practice, images are often sampled in a square array—that is, $M = N$. We use the following formulae:

$$F(u,v) = \frac{1}{N} \sum_{x=0}^{N-1} \sum_{y=0}^{N-1} f(x,y) \exp\left[-j2\pi\left(\frac{ux+vy}{N}\right)\right], \tag{3.3}$$

for $u,v = 0,1,2,\ldots,N-1$, and

$$f(x,y) = \frac{1}{N} \sum_{u=0}^{N-1} \sum_{v=0}^{N-1} F(u,v) \exp\left[j2\pi\left(\frac{ux+vy}{N}\right)\right], \tag{3.4}$$

for $x,y = 0,1,2,\ldots,N-1$. The DFT can be used for phase modulation between the watermark image and its carrier, as well as for dividing the image into perceptual bands to record the watermark. The DFT uses phase modulation instead of magnitude components to hide messages since phase modulation has less visual effect. Furthermore, phase modulation is more robust against noise attack.

The number of complex multiplications and additions required to implement the DFT is proportional to N^2. Its calculation is usually performed using a method known as the *fast Fourier transform* [2], whose decomposition can make the number of multiplication and addition operations proportional to $N \log_2 N$.

3.3 DISCRETE COSINE TRANSFORM

The Fourier series was originally motivated by the problem of heat conduction, and later found a vast number of applications as well as providing a basis for other transforms, such as discrete cosine transform (DCT). Many video and image compression algorithms apply the DCT to transform an image to the frequency domain and perform quantization for data compression. This helps separate an image into parts (or spectral subbands) of hierarchical importance (with respect to the image's visual quality). A well-known JPEG technology uses the DCT to compress images.

The Fourier transform kernel is complex valued. The DCT is obtained by using only a real part of the Fourier complex kernel. If $f(x, y)$ denotes an image in spatial domain and $F(u, v)$ denotes an image in frequency domain, the general equation for a 2D DCT is

$$F(u,v) = C(u)C(v) \sum_{x=0}^{N-1} \sum_{y=0}^{N-1} f(x,y) \cos\left(\frac{(2x+1)u\pi}{2N}\right) \cos\left(\frac{(2y+1)v\pi}{2N}\right), \quad (3.5)$$

where if $u = v = 0$, $C(u) = C(v) = \sqrt{\frac{1}{N}}$; otherwise, $C(u) = C(v) = \sqrt{\frac{2}{N}}$.

The inverse DCT can be represented as

$$f(x,y) = \sum_{u=0}^{N-1} \sum_{v=0}^{N-1} C(u)C(v)F(u,v) \cos\left(\frac{(2x+1)u\pi}{2N}\right) \cos\left(\frac{(2y+1)v\pi}{2N}\right). \quad (3.6)$$

A more convenient method for expressing the 2D DCT is with matrix products as $F = MfM^T$, and its inverse DCT is $f = M^T FM$, where F and f are 8×8 data matrices, and M is the matrix as

$$M =$$

$$
\begin{bmatrix}
\frac{1}{\sqrt{8}} & \frac{1}{\sqrt{8}} & \frac{1}{\sqrt{8}} & \frac{1}{\sqrt{8}} & \frac{1}{\sqrt{8}} & \frac{1}{\sqrt{8}} & \frac{1}{\sqrt{8}} & \frac{1}{\sqrt{8}} \\
\frac{1}{2}\cos\frac{1}{16}\pi & \frac{1}{2}\cos\frac{3}{16}\pi & \frac{1}{2}\cos\frac{5}{16}\pi & \frac{1}{2}\cos\frac{7}{16}\pi & \frac{1}{2}\cos\frac{9}{16}\pi & \frac{1}{2}\cos\frac{11}{16}\pi & \frac{1}{2}\cos\frac{13}{16}\pi & \frac{1}{2}\cos\frac{15}{16}\pi \\
\frac{1}{2}\cos\frac{2}{16}\pi & \frac{1}{2}\cos\frac{6}{16}\pi & \frac{1}{2}\cos\frac{10}{16}\pi & \frac{1}{2}\cos\frac{14}{16}\pi & \frac{1}{2}\cos\frac{18}{16}\pi & \frac{1}{2}\cos\frac{22}{16}\pi & \frac{1}{2}\cos\frac{26}{16}\pi & \frac{1}{2}\cos\frac{30}{16}\pi \\
\frac{1}{2}\cos\frac{3}{16}\pi & \frac{1}{2}\cos\frac{9}{16}\pi & \frac{1}{2}\cos\frac{15}{16}\pi & \frac{1}{2}\cos\frac{21}{16}\pi & \frac{1}{2}\cos\frac{27}{16}\pi & \frac{1}{2}\cos\frac{33}{16}\pi & \frac{1}{2}\cos\frac{39}{16}\pi & \frac{1}{2}\cos\frac{45}{16}\pi \\
\frac{1}{2}\cos\frac{4}{16}\pi & \frac{1}{2}\cos\frac{12}{16}\pi & \frac{1}{2}\cos\frac{20}{16}\pi & \frac{1}{2}\cos\frac{28}{16}\pi & \frac{1}{2}\cos\frac{36}{16}\pi & \frac{1}{2}\cos\frac{44}{16}\pi & \frac{1}{2}\cos\frac{52}{16}\pi & \frac{1}{2}\cos\frac{60}{16}\pi \\
\frac{1}{2}\cos\frac{5}{16}\pi & \frac{1}{2}\cos\frac{15}{16}\pi & \frac{1}{2}\cos\frac{25}{16}\pi & \frac{1}{2}\cos\frac{35}{16}\pi & \frac{1}{2}\cos\frac{45}{16}\pi & \frac{1}{2}\cos\frac{55}{16}\pi & \frac{1}{2}\cos\frac{65}{16}\pi & \frac{1}{2}\cos\frac{75}{16}\pi \\
\frac{1}{2}\cos\frac{6}{16}\pi & \frac{1}{2}\cos\frac{18}{16}\pi & \frac{1}{2}\cos\frac{30}{16}\pi & \frac{1}{2}\cos\frac{42}{16}\pi & \frac{1}{2}\cos\frac{54}{16}\pi & \frac{1}{2}\cos\frac{66}{16}\pi & \frac{1}{2}\cos\frac{78}{16}\pi & \frac{1}{2}\cos\frac{90}{16}\pi \\
\frac{1}{2}\cos\frac{7}{16}\pi & \frac{1}{2}\cos\frac{21}{16}\pi & \frac{1}{2}\cos\frac{35}{16}\pi & \frac{1}{2}\cos\frac{49}{16}\pi & \frac{1}{2}\cos\frac{63}{16}\pi & \frac{1}{2}\cos\frac{77}{16}\pi & \frac{1}{2}\cos\frac{91}{16}\pi & \frac{1}{2}\cos\frac{105}{16}\pi
\end{bmatrix}
$$

$$(3.7)$$

The obtained DCT coefficients indicate the correlation between the original 8×8 block and the respective DCT basis image. These coefficients represent the amplitudes of all cosine waves that are used to synthesize the original signal in the inverse process. Fourier cosine transform inherits many properties from Fourier transform; and there are other applications of DCT that can be noted [3, 4].

3.4 DISCRETE WAVELET TRANSFORM

Discrete wavelet transform (DWT) is also a simple and fast transformation approach that translates an image from spatial domain to frequency domain. Unlike DFT and DCT, which represent a signal either in spatial domain or in frequency domain, DWT is able to provide a representation for both spatial and frequency interpretations simultaneously. It is used in JPEG 2000 compression, and has become increasingly popular.

Wavelets are functions that integrate to zero waving above and below the x axis. Like sines and cosines in the Fourier transform, wavelets are used as the base functions for signal and image representation. Such base functions are obtained by dilating and translating a *mother wavelet* $\psi(x)$ by amounts s and τ, respectively:

$$\Psi_{\tau,s}(x) = \left\{ \psi\left(\frac{x-\tau}{s}\right), (\tau,s) \in R \times R^+ \right\}. \tag{3.8}$$

The translation and dilation allow the wavelet transform to be localized in time and frequency. Also, wavelet basis functions can represent functions with discontinuities and spikes in a more compact way than can sines and cosines.

CWT can be defined as

$$cwt_\psi(\tau,s) = \frac{1}{\sqrt{|s|}} \int x(t) \Psi^*_{\tau,s}(t) dt, \tag{3.9}$$

where $\Psi^*_{\tau,s}$ is the complex conjugate of $\Psi_{\tau,s}$ and $x(t)$ is the input signal defined in the time domain.

Inverse CWT can be obtained as

$$x(t) = \frac{1}{C_\psi^2} \int_s \int_\tau cwt_\psi(\tau,s) \frac{1}{s^2} \Psi_{\tau,s}(t) d\tau ds, \tag{3.10}$$

where C_ψ is a constant and depends on the wavelet used.

To discretize the CWT, the simplest case is the uniform sampling of the time-frequency plane. However, the sampling could be more efficient by using the Nyquist rule:

$$N_2 = \frac{s_1}{s_2} N_1, \tag{3.11}$$

where N_1 and N_2 denote the number of samples at scales s_1 and s_2, respectively, and $s_2 > s_1$. This rule means that at higher scales (lower frequencies), the number of samples can be decreased. The sampling rate obtained is the minimum rate that allows the original signal to be reconstructed from a discrete set of samples. A dyadic scale satisfies the Nyquist rule by discretizing the scale parameter into a logarithmic series, and the time parameter is then discretized with respect to the corresponding scale parameters. The following equations set the translation and dilation to the dyadic scale with logarithmic series of base 2 for $\psi_{k,j}$: $\tau = k2^j$, $s = 2^j$.

We can view these coefficients as filters that are classified into two types. One set, H, works as a low-pass filter, and the other, G, as a high-pass filter. These two types of coefficients are called *quadrature mirror filters* and are used in pyramidal algorithms.

For a 2D signal, the 2D wavelet transform can be decomposed by using the combination of 1D wavelet transforms. The 1D transform can be applied individually to each of the dimensions of the image. By using quadrature mirror filters we can decompose an $n \times n$ image I into the wavelet coefficients, as below. Filters H and G are applied on the rows of an image, splitting the image into two subimages of dimensions $n/2 \times n$ (half the columns) each. One of these subimages, $H_r I$ (where the subscript r denotes row), contains the low-pass information; the other, $G_r I$, contains the high-pass information. Next, the filters H and G are applied to the columns of both subimages. Finally, four subimages with dimensions $n/2 \times n/2$ are obtained. Subimages $H_c H_r I$, $G_c H_r I$, $H_c G_r I$, and $G_c G_r I$ (where the subscript c denotes *column*) contain the low-low, high-low, low-high, and high-high passes, respectively. Figure 3.2 illustrates this decomposition. The same procedures are applied iteratively to the subimage containing the most low-band information until the subimage's size reaches 1×1. Therefore, the initial dimensions of the image are required to be powers of two.

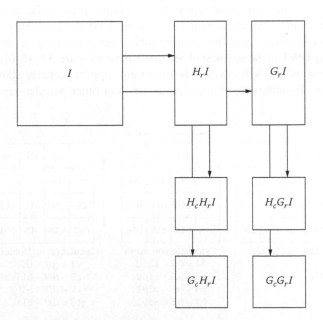

FIGURE 3.2 The wavelet decomposition of an image.

FIGURE 3.3 The three-level wavelet decomposition of the Lena image.

In practice, it is not necessary to carry out all the possible decompositions until the size of 1×1 is reached. Usually, just a few levels are sufficient because most of the object features can be extracted from them. Figure 3.3 shows the Lena image decomposed in three levels. Each of the resulting subimages is known as a subband.

The Haar DWT is the simplest of all wavelets (see Figure 3.4) [5, 6]. Haar wavelets are orthogonal and symmetric. The minimum support property allows arbitrary grid intervals. Boundary conditions are easier than other wavelet-based methods.

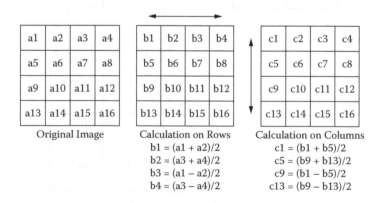

a1	a2	a3	a4
a5	a6	a7	a8
a9	a10	a11	a12
a13	a14	a15	a16

Original Image

b1	b2	b3	b4
b5	b6	b7	b8
b9	b10	b11	b12
b13	b14	b15	b16

Calculation on Rows
$b1 = (a1 + a2)/2$
$b2 = (a3 + a4)/2$
$b3 = (a1 - a2)/2$
$b4 = (a3 - a4)/2$

c1	c2	c3	c4
c5	c6	c7	c8
c9	c10	c11	c12
c13	c14	c15	c16

Calculation on Columns
$c1 = (b1 + b5)/2$
$c5 = (b9 + b13)/2$
$c9 = (b1 - b5)/2$
$c13 = (b9 - b13)/2$

FIGURE 3.4 An example of Haar DWT.

The Haar DWT works very well to detect the characteristics like edges and corners. Figure 3.4 shows an example of DWT of size 4×4.

3.5   RANDOM SEQUENCE GENERATION

After a watermark is embedded into a host image, if we crop or damage the water-marked image, the recipient would not be able to extract the complete watermark. The extracted watermark will be distorted and some important information may be lost. To overcome the loss of information, we can rearrange the pixel sequence of a watermark via a random sequence generator [7].

For example, in Figure 3.5 we use the 14-bit random sequence generator to relo-cate the pixel order of a watermark of 128×128 by moving the 13th bit to the rear (bit 0) of the whole sequence. A 2D watermark is first raster-scanned into a 1D image of size 16,384 that can be stored in a 1D array with a 13-bit index. Let (x, y) denote the column and row numbers respectively, and let the upper left corner of the 2D image be (0,0). For example, a pixel located at (72,101) has the index 13000 (i.e., $128 \times 101 + 72$). Its binary code is 11001011001000 in the 1D array. After the 14-bit random sequence generator, the new index becomes 9617 (i.e., 10010110010001) in the 1D array. The corresponding location in the 2D image is (17,75), as shown in Figure 3.6.

Another random sequence to spread the watermark is calculated using the secret key. Each value of the secrete key is used as a seed to generate a short random sequence. When concatenating the short random sequences and randomly permuting the resulted sequence, we can obtain the pseudorandom sequence that will be used to spread the watermark.

Another random sequence uses the signature noise. Two keys, the owner key (the one known to the public) and the local key (the one calculated from image samples by a hash function), are used to generate the noise sequence. The local key and the owner key are combined to generate the random sequence, which results in a sequence with zero mean and an autocorrelation function close to an impulse. We can obtain a normal distribution by averaging each set of eight iterations of the generator. Eight have been selected because the division by a power of two is equivalent to a shift operator, which only consumes one central processing unit (CPU) cycle.

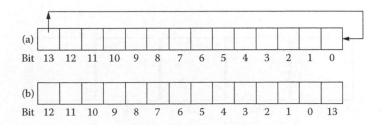

FIGURE 3.5 An example of random sequence generation.

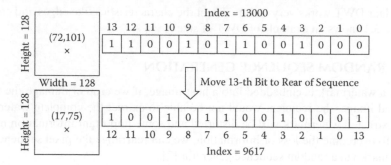

FIGURE 3.6 An example of pixel relocation by random sequence generation.

3.6 THE CHAOTIC MAP

Although the random sequence generation discussed in the previous section can rearrange the pixel permutation of a watermark, it does not spread the neighboring pixels into dispersed locations sufficiently. As a result, the relocated pixels are still close to one another. For example, Figure 3.7(a) shows an image containing a short line, and 3.7(b)–3.7(e) show the results after moving the 13th, 12th, 11th, and 10th bits to the rear location, respectively. They are still perceived as lines. This happens because the bit-relocation approach does not change the spatial relationship among pixels. Table 3.1 illustrates a line of 11 pixels. We use random sequence generation to shift the 14th bit to the rear location. After pixel relocation, the distance between each pair of connected pixels is changed from 1 to 2. As an alternative, we can spread the pixels by swapping the 13th bit with the 0th bit. However, this will separate the pixels into two subgroups.

In order to spread the neighboring pixels into largely dispersed locations, we often use the chaotic map. In mathematics and physics, chaos theory deals with the behavior of certain nonlinear dynamic systems that exhibit a phenomenon under certain conditions known as *chaos* [8, 9], which adopts the Shannon requirement on diffusion and confusion [10]. Its characteristics are well taken in robust digital watermarking to enhance the security. The most critical aspects of chaotic functions are its utmost sensitivity to initial conditions and its outspreading behavior over the entire space. Therefore, chaotic maps are very useful in watermarking and encryption.

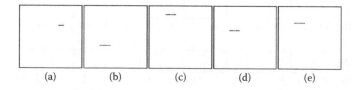

FIGURE 3.7 The pixel information in four different strategies of random sequence generation.

TABLE 3.1
Pixel Information after Random Sequence Generation

Original Index	Binary Format	Original 2D Location	Move the 13th Bit to the Rear	New Index	New 2D Location
5200	01010001010000	(80, 40)	10100010100000	10400	(32, 81)
5201	01010001010001	(81, 40)	10100010100010	10402	(34, 81)
5202	01010001010010	(82, 40)	10100010100100	10404	(36, 81)
5203	01010001010011	(83, 40)	10100010100110	10406	(38, 81)
5204	01010001010100	(84, 40)	10100010101000	10408	(40, 81)
5205	01010001010101	(85, 40)	10100010101010	10410	(42, 81)
5206	01010001010110	(86, 40)	10100010101100	10412	(44, 81)
5207	01010001010111	(87, 40)	10100010101110	10414	(46, 81)
5208	01010001011000	(88, 40)	10100010110000	10416	(48, 81)
5209	01010001011001	(89, 40)	10100010110010	10418	(50, 81)
5210	01010001011010	(90, 40)	1010001011010	10420	(52, 81)

The iterative values generated from chaotic functions are completely random in nature and never converge, although they are limited within a range.

The logistic map is one of the simplest chaotic maps [11], expressed as

$$x_{n+1} = r \cdot x_n(1 - x_n),$$ (3.12)

where x takes values in the interval $(0, 1)$. If parameter r is between 0 and 3, the logistic map will converge to a particular value after some iterations. As r continues to increase, the curves' bifurcations become faster and when r is greater than 3.57, the curves complete chaotic periodicity. Finally, if r is in the interval $(3.9, 4)$, the chaotic values are in the range of $(0, 1)$.

The chaotic map for changing pixel (x, y) to (x', y') can be represented by

$$\begin{bmatrix} x' \\ y' \end{bmatrix} = \begin{bmatrix} 1 & 1 \\ l & l+1 \end{bmatrix} \begin{bmatrix} x \\ y \end{bmatrix} \bmod N$$ (3.13)

where $\det\left(\begin{bmatrix} 1 & 1 \\ l & l+1 \end{bmatrix}\right) = 1$ or -1, and l and N respectively denote an integer and the width of a square image. The chaotic map is applied to the new image iteratively by equation 3.13.

Figure 3.8 illustrates the chaotic map, as applied (in Figure 3.8(a)) on an image of size 101×101, which contains a short horizontal line. After using equation 3.13 with $l = 2$ for four iterations, the results are shown in Figures 3.8(b)–3.8(e). We observe that the neighboring pixels are now scattered everywhere around the image. Table 3.2 lists the coordinate change of each pixel using the chaotic map.

Figure 3.9(a) shows a Lena image of size 101×101. Figures 3.9(b)1–3.9(b)17 are the iterative results of applying equation 3.13 with $l = 2$. Note that the resulting image is exactly the same as the original image after 17 iterations.

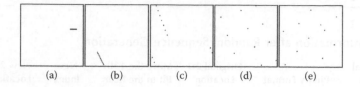

(a)   (b)   (c)   (d)   (e)

FIGURE 3.8 The pixel information in the first four iterations of the chaotic map.

TABLE 3.2

The Coordinate Changes of Each Pixel during the Chaotic Map Procedure

Original 2D Location	2D Location in 1st Iteration	2D Location in 2nd Iteration	2D Location in 3rd Iteration	2D Location in 4th Iteration
(80, 40)	(19, 78)	(97, 70)	(66, 0)	(66, 31)
(81, 40)	(20, 80)	(100, 78)	(77, 30)	(6, 42)
(82, 40)	(21, 82)	(2, 86)	(88, 60)	(47, 53)
(83, 40)	(22, 84)	(5, 94)	(99, 90)	(88, 64)
(84, 40)	(23, 86)	(8, 1)	(9, 19)	(27, 75)
(85, 40)	(24, 88)	(11, 9)	(20, 49)	(69, 86)
(86, 40)	(25, 90)	(14, 17)	(31, 79)	(9, 97)
(87, 40)	(26, 92)	(17, 25)	(42, 8)	(50, 7)
(88, 40)	(27, 94)	(20, 33)	(53, 38)	(91, 18)
(89, 40)	(28, 96)	(23, 41)	(64, 68)	(31, 29)
(90, 40)	(29, 98)	(26, 49)	(75, 98)	(72, 40)

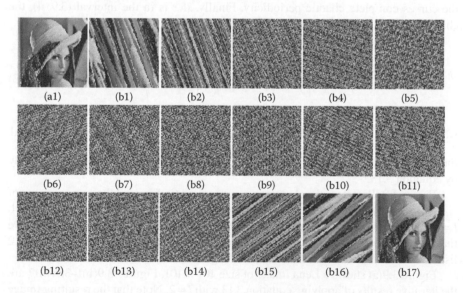

(a1)   (b1)   (b2)   (b3)   (b4)   (b5)

(b6)   (b7)   (b8)   (b9)   (b10)   (b11)

(b12)   (b13)   (b14)   (b15)   (b16)   (b17)

FIGURE 3.9 An example of performing chaotic mapping on the Lena image.

3.7 ERROR CORRECTION CODE

Error correction code (ECC) is another technology used to improve robustness in digital watermarking. It was designed to represent a sequence of numbers such that any errors occurring can be detected and corrected (within certain limitations) [12]. It is often used in coding theory to allow data that is being read or transmitted to be checked for errors and, when necessary, corrected. This is different from parity checking, which only detects errors. Nowadays, ECC is often designed into data storage and transmission hardware.

In digital watermarking, hidden data can be partially or completely destroyed due to some attacks, such as those from noise, smooth filters, and JPEG compression. In order to detect and correct the error message, specific coding rules and additional information are required. That is, after translating a watermark image into a bitstream, we conduct reconstruction by specific rules and by adding some information, so a recipient is able to detect whether the bitstream (i.e., the watermark) is correct or incorrect. If the recipient detects some bits within the bitstream that are incorrect, she is able to correct them based on the additional information. Some commonly used ECCs are binary Golay code, Bose Ray–Chaudhuri Hocquenghem code, convolutional code, low-density parity-check code, Hamming code, Reed-Solomon code, Reed-Muller code, and turbo code.

The basic coding approach is to translate the symbol into a binary code. Figure 3.10 shows a coding example in which 32 symbols (0–9 and A–V) are coded into a 5-bit binary stream. The information rate in equation 3.14 is used to evaluate the coding efficiency, where A is the code word and n is the length of the coded bitstream:

$$Information\ Rate = \frac{\log_2 A}{n} \qquad (3.14)$$

The range of information rate is between 0 and 1. A higher value of information rate indicates a higher coding efficiency.

For example, in Figure 3.10, there are 32 symbols ($A = 32$), and the length of the coded bitstream is 5 ($n = 5$). Therefore, the information rate is 1, meaning a perfect coding efficiency—that is, there is no waste in the coding.

0	00000	E	01101
1	00001	F	01110
2	00010	G	01111
⋮	⋮	⋮	⋮
9 ⟹	01001	S ⟹	11100
A	01010	T	11101
B	01011	U	11110
C	01100	V	11111

FIGURE 3.10 A coding example of 32 symbols.

2-bit Coding	Receiver Obtain....		
4 Cases: {00, 01, 10, 11}	00	No Error	00
Only Use Two Cases	10 ⇨	Error	⇨ ??
{00, ~~01, 10~~, 11}	01	Error	??

FIGURE 3.11 An example of detecting correctness of the obtained message.

If the watermarked image is under an attack, the hidden data may be destroyed and the recipient cannot identify the correct message. For example, if a recipient extracts 00010 from a watermarked image, he cannot be sure whether this bitstream indicates 2 (if no bit error) or A (if the 3rd bit was mistakenly changed from 1 to 0). We can provide specific coding rules, so a recipient can check the code to determine whether the message from the extracted watermark is altered. Figure 3.11 shows an example in which a recipient is able to know the correctness of the obtained message, where a 2-bit code containing four possibilities is given. The coding rule is defined as choosing only two cases—00 and 11—for coding. If 01 or 10 is obtained, the recipient could know that the code is wrong. However, the recipient cannot be sure whether the correct message should be 00 or 11, since the minimum Hamming distance between the obtained message (10/01) and the correct code (00/11) is the same as 1. Note that the Hamming distance between two bitstreams of same length is equal to the number of positions for which the corresponding symbols are different.

In order to correct the coding errors, additional information is required. For example, redundant codes are usually used to correct the errors. Figure 3.12 shows an example of the redundant code for correcting the error, where two copies of each bit are duplicated to form a 3-bit stream space. Therefore, a recipient can receive a total of eight possible cases within the 3-bit stream space. If the recipient obtains 001, 010, or 100, he knows there is a bit error and corrects it to be 000, so the message is 0. Similarly, if the recipient obtains 110, 101, or 011, he/she can correct it and determines that the message is 1.

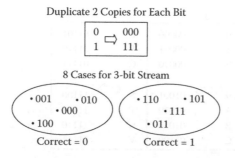

Duplicate 2 Copies for Each Bit

| 0 | ⇨ | 000 |
| 1 | | 111 |

8 Cases for 3-bit Stream

• 001 • 010	• 110 • 101
• 000	• 111
• 100	• 011
Correct = 0	Correct = 1

FIGURE 3.12 An example of redundant coding for error correction.

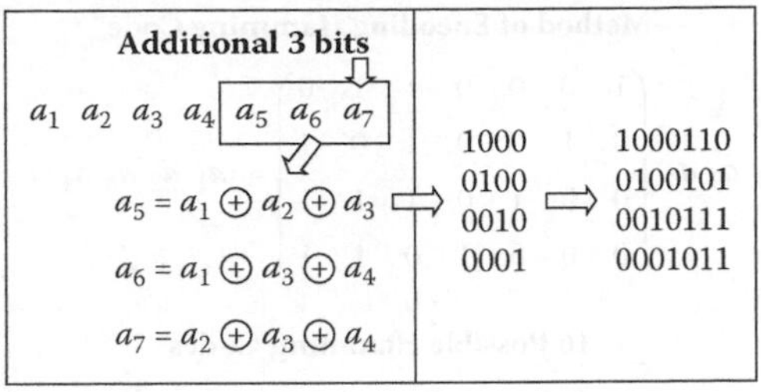

FIGURE 3.13 The rule of adding 3 bits for Hamming code (7, 4).

The Hamming code is a famous error-correcting code. For the remainder of this section, the Hamming code (7, 4) is introduced to check every 4 bits by adding 3 additional bits. The idea of the Hamming code is shown in Figure 3.13, where 3 additional bits are added and their values are related to the original 4 bits. For example, bits 5, 6, and 7 are respectively derived by ($bit\ 1 \oplus bit\ 2 \oplus bit\ 3$), ($bit\ 1 \oplus bit\ 3 \oplus bit\ 4$), and ($bit\ 2 \oplus bit\ 3 \oplus bit\ 4$), where $\oplus$ denotes the *exclusive or* (*XOR*) Boolean operator. Therefore, a bitstream 1000 will become 1000110.

Hamming matrices, as shown in Figure 3.14, including the code-generation matrix and parity-check matrix, can be used to encode a 4-bit stream and detect errors of the obtained 7-bit coded stream. The method of encoding 16 possible Hamming codes by translating 4-bit streams into 7-bit coded streams is shown in Figure 3.15.

An example of error correction is shown in Figure 3.16. By multiplying the obtained code stream and the parity-check matrix, the syndrome vector can be obtained to indicate whether an error is occurred and, if so, for which codeword bit. In Figure 3.16(a), the syndrome vector is (0,0,0) indicating there is no error. In Figure 3.16(b), the syndrome vector is (1,0,1), corresponding to the second row of the parity-check matrix. Therefore, the recipient will know that the second bit of the obtained bitstream is incorrect.

Hamming Matrices

Code Generation Matrix   Parity-Check Matrix

$$G = \begin{pmatrix} 1 & 0 & 0 & 0 & 1 & 1 & 0 \\ 0 & 1 & 0 & 0 & 1 & 0 & 1 \\ 0 & 0 & 1 & 0 & 1 & 1 & 1 \\ 0 & 0 & 0 & 1 & 0 & 1 & 1 \end{pmatrix} \qquad H = \begin{pmatrix} 1 & 1 & 0 \\ 1 & 0 & 1 \\ 1 & 1 & 1 \\ 0 & 1 & 1 \\ 1 & 0 & 0 \\ 0 & 1 & 0 \\ 0 & 0 & 1 \end{pmatrix}$$

FIGURE 3.14 The Hamming matrices.

Method of Encoding Hamming Code

$$(a_1 \ a_2 \ a_3 \ a_4) \begin{pmatrix} 1 & 0 & 0 & 0 & 1 & 1 & 0 \\ 0 & 1 & 0 & 0 & 1 & 0 & 1 \\ 0 & 0 & 1 & 0 & 1 & 1 & 1 \\ 0 & 0 & 0 & 1 & 0 & 1 & 1 \end{pmatrix} = (a_1 \ a_2 \ a_3 \ a_4 \ a_5 \ a_6 \ a_7)$$

16 Possible Hamming Codes

0000000	0001011	0010111	0011100
0100101	0101110	0110010	0111001
1000110	1001101	1010001	1011000
1100011	1101001	1110100	1111111

FIGURE 3.15 The 16 possible Hamming codes.

3.8 SET PARTITIONING IN HIERARCHICAL TREE

Set partitioning in hierarchical tree (SPIHT), introduced by Said and Pearlman [13], is a zerotree structure for image coding based on DWT. The first zerotree structure, called an embedded zerotree wavelet (EZW), was published by Shapiro in 1993 [14]. The SPIHT coding uses a bit allocation strategy to produce a progressively embedded scalable bitstream.

A primary feature of the SPIHT coding is that it converts the image into a sequence of bits based on the priority of transformed coefficients of an image. The magnitude of each coefficient is taken to determine whether it is important. The larger absolute value a coefficient has, the more important it is. Therefore, after the SPIHT coding, the length of the coded string may be varied according to the quality requirement. In general, when a decoder reconstructs the image based on the coded string, a higher peak signal-to-noise ratio is achieved. The reconstructed image is similar to the original one if it can have the coded string as long as possible. It is no doubt that a decoder can reconstruct the original image if the completed coded string is available.

Basically, the SPIHT coding algorithm will encode the significant coefficients iteratively. In each iteration, if the absolute value of a coefficient is larger than the predefined threshold, the coefficient will be coded and its value (i.e., magnitude)

$$(1110110) \begin{pmatrix} 1 & 1 & 0 \\ 1 & 0 & 1 \\ 1 & 1 & 1 \\ 0 & 1 & 1 \\ 1 & 0 & 0 \\ 0 & 1 & 0 \\ 0 & 0 & 1 \end{pmatrix} = (000) \qquad (1111010) \begin{pmatrix} 1 & 1 & 0 \\ 1 & 0 & 1 \\ 1 & 1 & 1 \\ 0 & 1 & 1 \\ 1 & 0 & 0 \\ 0 & 1 & 0 \\ 0 & 0 & 1 \end{pmatrix} = (101)$$

(a)   (b)

FIGURE 3.16 An example of error correction based on Hamming code (7, 4).

will be decreased by a refinement procedure. The threshold is also decreased in each iteration. Following are the relative terms in the SPIHT coding algorithm:

n: The largest factor determined based on the largest absolute coefficient by

$$n = [\log_2[\max(|\,c(i,j)\,|)]] \qquad (3.15)$$

H: A set of initial coefficients that are usually the low-frequency coefficients. For example, the coefficients (0,0), (0,1), (1,0), and (1,1) in Figure 3.17.

LSP: List of significant partitioning coefficients. The significant coefficient determined in each iteration will be added into the LSP. It is set to be empty upon initialization.

LIP: List of insignificant partitioning coefficients. Initially, all the coefficients in set *H* are considered as insignificant and added into the LIP.

LIS: List of the initial unprocessed coefficients.

O(i, j): Set of all offspring of coefficient (*i*,*j*). Excepting the highest and lowest pyramid levels in the quad-tree structure, *O(i,j)* can be expressed by

$$O(i, j) = \{(2i, 2j), (2i, 2j+1), (2i+1, 2j), (2i+1, 2j+1)\} \qquad (3.16)$$

For example, $O(3, 3) = \{(6, 6),(7, 6),(6, 7),$ and $(7, 7)\}$ in Figure 3.17.

D(i, j): Set of all descendants of coefficient (*i*, *j*).

L(i, j): Set of all descendants of coefficient (*i*, *j*) except *O(i, j)*. That is, $L(i, j) = D(i, j) - O(i, j)$.

c(i, j): The value of coefficient (*i*, *j*) in the transformed image.

$S_n(i, j):$ The significant record of coefficient (*i*,*j*). It is determined by

$$S_n(i,j) = \begin{cases} 1, & if \ |c(i,j)| \geq 2^n \\ 0, & otherwise \end{cases} \qquad (3.17)$$

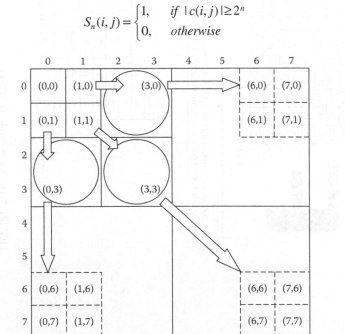

FIGURE 3.17 An example of quad-tree structure.

(a) Original 8*8 Image

192	192	192	192	192	192	192	192
192	16	16	16	16	16	192	192
192	16	192	192	192	192	192	192
192	16	16	16	16	192	192	192
192	16	192	192	16	192	192	192
192	16	192	192	16	192	192	192
192	16	192	192	192	192	192	192
192	192	192	192	192	192	192	192

(b) Transformed Image by DWT

139.75	-24.75	11	-11	44	0	0	-44
-2.75	275	11	-22	88	-88	0	0
11	-33	11	-11	88	-44	0	0
-33	-22	11	-22	44	-44	0	0
44	88	88	44	-44	0	0	44
0	0	0	0	0	0	0	0
0	-44	-88	0	0	44	0	0
-44	-44	0	0	44	-44	0	0

110000001
0000100001000011100111001110011100110111000011000100000
0000100010101001000011111001010010011101010101001001100001011001011110011111110000000000000000

(c) The Bitstreams by SPIHT

(d) The Reconstructed Image

138.25	0	0	40	0	-40
0	0	92.5	92.5	0	0
0	-40	0	92.5	-40	0
-40	0	0	40	-40	0
40	92.5	92.5	-40	40	40
0	0	0	0	0	0
0	-40	-92.5	0	40	0
-40	-40	0	40	-40	0

FIGURE 3.18 An example of SPIHT.

THE SPIHT CODING ALGORITHM

1. Initialization: Determine n by equation 3.15; set LSP and LIS to be empty; add all elements in H into LIP.
2. Sorting Pass
   2.1. For each entry (i, j) in LIP do
       output $S_n (i, j)$;
       if $S_n(i, j) = 1$ then move (i, j) into LSP and out the sign of $c(i, j)$;
   2.2. For each entry (i, j) in LIS do
       2.2.1 if the entry is of type A then
           output $S_n(D(i, j))$;
           if $S_n(D(I, j)) = 1$ then
             for each $(k, l) \in O\ (i, j)$ do
               output $s_n\ (k, l)$;
               if $s_n\ (k, l) = 1$ then add (k, l) into LSP and out the sign of $c(k, l)$;
               if $S_n\ (k, l) = 0$ then add (k, l) to the end of LIP;
             if $L(i, j)! = 0$ then move (i, j) to the end of the LIS, as an entry of type B,
             or else remove entry (i, j) from the LIS;
       2.2.2. if the entry is of type B then
           output $S_n\ (L(i, j))$;
           if $S_n\ (L(i, j)) = 1$ then
             add each $(k, l) \in O(i, j)$ to the end of the LIS as an entry of type A;
             remove (i, j) from the LIS;
3. Refinement Pass: for each entry (i, j) in the LSP, except those included in the last sorting pass, output the nth most significant bit of $c(i, j)$;
4. Update: Decrease n by 1 and go to step 2.

Figure 3.18 shows an example of SPIHT, where Figure 3.18(a) is the original 8×8 image. We obtain the transformed image as Figure 3.18(b) by DWT. Figure 3.18(c) lists the bitstreams generated by applying SPIHT three times. We reconstruct the image by using these bitstreams, and the result is shown in Figure 3.18(d).

REFERENCES

[1] Johnson, N., and S. Jajodia, S. "Exploring Steganography: Seeing the Unseen." *IEEE Computer* 31 (1998): 26.
[2] Brigham, E. *The Fast Fourier Transform and Its Applications.* Upper Saddle River, NJ: Prentice-Hall, 1988.
[3] Rao, K., and P. Yip. *Discrete Cosine Transform: Algorithms, Advantages, and Applications.* Academic Press, 1990.
[4] Lin, S. D., and C.-F. Chen. "A Robust DCT-Based Watermarking for Copyright Protection." *IEEE Trans. Consumer Electronics* 46 (2000): 415.
[5] Falkowski, B. J. "Forward and Inverse Transformations between Haar Wavelet and Arithmetic Functions." *Electronics Letters* 34 (1998): 1084.

[6] Grochenig, K., and W. R. Madych. "Multiresolution Analysis, Haar Bases, and Self-Similar Tilings of R^n." *IEEE Trans. Information Theory* 38 (1992): 556.

[7] Knuth, D. *The Art of Computer Programming*. Vol. 2. Reading, MA: Addison-Wesley, 1981.

[8] Ott, E. *Chaos in Dynamical Systems*. New York: Cambridge University Press, 2002.

[9] Zhao, D., G. Chen, and W. Liu. "A Chaos-Based Robust Wavelet-Domain Watermarking Algorithm." *Chaos, Solitons and Fractals* 22 (2004): 47.

[10] Schmitz, R. "Use of Chaotic Dynamical Systems in Cryptography." *J. Franklin Institute* 338 (2001): 429.

[11] Devaney, R. *An Introduction to Chaotic Dynamical Systems*. 2d ed. Redwood City, CA: Addison-Wesley, 1989.

[12] Peterson, W., and E. Weldon. *Error-Correcting Codes*. Boston: MIT Press, 1972.

[13] Said, A., and W. A. Pearlman. "A New, Fast, and Efficient Image Code Based on Set Partitioning in Hierarchical Trees." *IEEE Trans. Circuits and Systems for Video Technology* 6 (1996): 243.

[14] Shapiro, J. "Embedded Image Coding Using Zerotrees of Wavelet Coefficients." *IEEE Trans. Signal Processing* 41 (1993): 3445.

4 Digital Watermarking Fundamentals

Digital watermarking is used for such information security as copyright protection, data authentication, broadcast monitoring, and covert communication. The watermark is embedded into a host image in such a way that the embedding-induced distortion is too small to be noticed. At the same time, the embedded watermark must be robust enough to withstand common degradations or deliberate attacks. Additionally, for given distortion and robustness levels, one would like to embed as much data as possible in a given host image. Podilchuk and Delp have presented a general framework for watermark embedding and detection/decoding [1], comparing the differences of various watermarking algorithms and applications in copyright protection, authentication, tamper detection, and data hiding.

This chapter introduces the fundamental digital watermarking techniques, including the spatial-domain and frequency-domain embedding approaches and fragile and robust watermarking techniques.

4.1 SPATIAL-DOMAIN WATERMARKING

Spatial-domain watermarking is the modifying of pixel values directly on the spatial domain of an image [2]. In general, spatial-domain watermarking schemes are simple and do not need the original image to extract the watermark. They also provide a better compromise among robustness, capacity, and imperceptibility. However, they have the disadvantage of not being robust against image-processing operations because the embedded watermark is not distributed around the entire image and the operations can thus easily destroy the watermark. The purpose of watermarking is to embed a secret message at the content level under the constraints of imperceptibility, security, and robustness against attacks. We could categorize most watermark embedding algorithms into substitution by a codebook element or additive embedding.

4.1.1 SUBSTITUTION WATERMARKING IN THE SPATIAL DOMAIN

Substitution watermarking in the spatial domain is the simplest watermarking algorithm [3, 4]. Basically, the embedding locations, such as the specific bits of all pixels, are predefined before watermark embedding. Once the recipient obtains the watermarked image, she knows the exact locations from which to extract the watermark. During the watermark embedding procedure, the watermark is first converted into a bitstream. Then, each bit of the bitstream is embedded into the specific bit of selected locations for the host image. Figure 4.1 shows an example of substitution watermarking in the spatial domain, in which there are three least-significant-bit (LSB) substitutions performed (i.e., embedding 1, 0, and 1 into 50, 50 and 48, respectively).

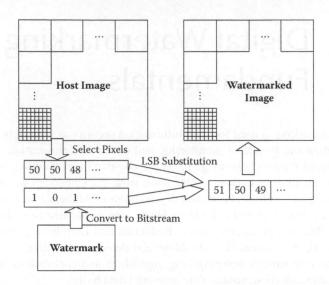

FIGURE 4.1 The watermark embedding procedure of substitution watermarking in the spatial domain.

During the watermark extraction procedure, the specific pixel locations of the watermarked image are already known. Then, each pixel value is converted into its binary format. Finally, the watermark is collected from the bit, where the watermark is embedded. Figure 4.2 shows an example of the watermark extracting procedure of substitution watermarking in the spatial domain.

In general, the watermark capacity of substitution watermarking in the spatial domain is larger than other watermarking approaches. Its maximum watermark capacity is eight times that of the host image. However, for the purpose of imperceptibility, a reasonable watermark capacity is three times that of the host image. If

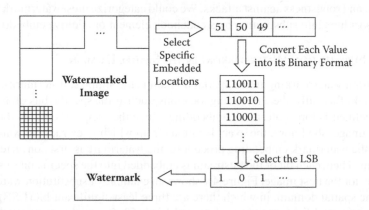

FIGURE 4.2 The watermark extracting procedure of the substitution watermarking in the spatial domain.

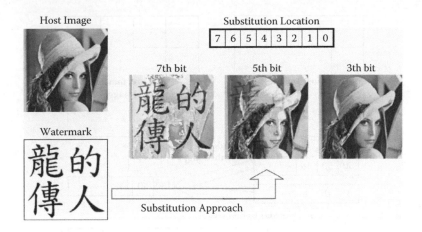

FIGURE 4.3 An example of embedding a watermark into different bits.

the watermark is embedded into the least three significant bits, human beings cannot distinguish the original image from the watermarked image. Figure 4.3 shows an example of embedding a watermark into different bits. It is clear that the water-marked image can be distinguished by human eyes if the watermark is embedded into the 7th and 5th bits. However, if the watermark is embedded into the 3rd bit, it would be difficult to tell whether there is a watermark hidden in the image.

Substitution watermarking is easy to implement. However, the embedded water-mark is not robust against collage or lossy compression attack. Therefore, some improved spatial-domain watermarking algorithms have been proposed to target robustness—for example, the hash function method and the bipolar M-sequence method [5, 6]. In contrast to robust watermarking, fragile watermarking can embed a breakable watermark into an image for the purpose of later detecting whether the image was altered.

4.1.2 ADDITIVE WATERMARKING IN THE SPATIAL DOMAIN

Different from the substitution approach, the additive watermarking approach does not consider specific bits of a pixel. Instead, it adds an amount of watermark value into a pixel to perform the embedding approach. If H is the original grayscale host image, W the binary watermark image, then $\{h(i,j)\}$ and $\{w(i,j)\}$ denote their respective pixels. We can embed W into H to become the watermarked image H^* as

$$h^*(i,j) = h(i,j) + a(i,j) \cdot w(i,j), \qquad (4.1)$$

where $\{a(i,j)\}$ denotes the scaling factor. Basically, the larger a(i,j) is, the more robust the watermarking algorithm is. Figure 4.4 shows an example of additive watermarking in the spatial domain, in which if the embedded watermark is 1, then a$(i,j) = 100$; other-wise, a$(i,j) = 0$. Therefore, after watermark embedding, the original values are changed from 50, 50, and 48 to 150, 50, and 148 using watermarks 1, 0, and 1, respectively.

It is obvious that if we have a large a(i,j), the watermarked image would be distorted, as shown in Figure 4.3. It is then difficult to achieve high imperceptibility.

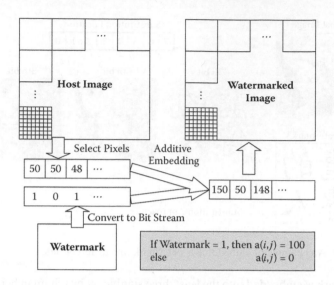

FIGURE 4.4 The watermark embedding procedure of additive watermarking in the spatial domain.

Two critical issues to consider in the additive watermarking are imperceptibility and the need of the original image to identify the embedded message.

In order to enhance imperceptibility, a big value is not embedded to a single pixel, but to a block of pixels [7, 8]. Figure 4.5 shows an example of the block-based additive watermarking. First, a 3×3 block is selected for embedding a watermark by adding 99. Second, the amount is divided by 9.

The second issue is the need of the original image in extracting the embedded watermark. When a recipient obtains the watermarked image, it is difficult to

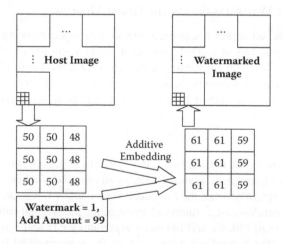

FIGURE 4.5 An example of block-based additive watermarking in the spatial domain.

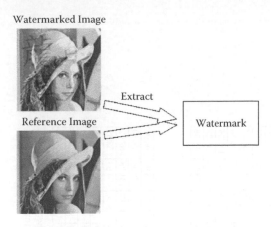

FIGURE 4.6 The reference image is required in the additive watermarking to extract the watermark.

determine the embedded message since he does not know the location of the block for embedding. Therefore, the reference (or original) image is usually required to extract the watermark, as shown in Figure 4.6.

4.2 FREQUENCY-DOMAIN WATERMARKING

In frequency- (or spectrum-) domain watermarking [9, 10], we can insert a watermark into frequency coefficients of the transformed image via *discrete Fourier transform* (DFT), *discrete cosine transform* (DCT), or *discrete wavelet transform* (DWT). Because the frequency transforms usually decorrelate the spatial relationships of pixels, the majority of the energy concentrates on the low-frequency components. When we embed the watermark into the low or middle frequencies, these changes will be distributed throughout the entire image. When image processing operations are applied on the watermarked image, they are affected less. Therefore, when compared to the spatial-domain watermarking method, the frequency-domain watermarking technique is relatively more robust.

This section describes the substitution and multiplicative watermarking methods in the frequency domain, and will introduce the watermarking scheme based on vector quantization. Finally, it will present a rounding error problem in frequency-domain methods.

4.2.1 SUBSTITUTION WATERMARKING IN THE FREQUENCY DOMAIN

The substitution watermarking scheme in the frequency domain is basically similar to that of the spatial domain except that the watermark is embedded into frequency coefficients of the transformed image. Figure 4.7 shows an example of embedding a 4-bit watermark into the frequency domain of an image. Figure 4.7(a) is an 8 × 8 grayscale host image, and Figure 4.7(b) is the transformed image by DCT. Figure 4.7(c) is a binary watermark, in which *0* and *1* are the embedded values and a hyphen (-) indicates no change in its position. We obtain Figure 4.7(d) by embedding 4.7(c) into 4.7(b) using the LSB substitution. Figure 4.8 shows the details of the LSB substitution approach.

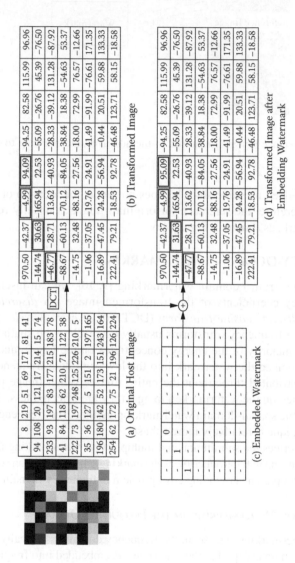

FIGURE 4.7 An example of embedding a 4-bit watermark in the frequency domain of an image.

LSB (Least Significant Bit) Substitution

W	Original Coeff.	Integer Part	Binary Format	Watermarked Binary	Watermarked Coeff.
1	−46.77	46	00101110	00101111	−47.77
1	30.63	30	00011110	00011111	31.63
0	−4.99	4	00000100	00000100	−4.99
1	94.09	94	01011110	01011111	94.09

FIGURE 4.8 Embedding a watermark into coefficients of the transformed image in Figure 4.7.

4.2.2 MULTIPLICATIVE WATERMARKING IN THE FREQUENCY DOMAIN

Frequency-domain embedding inserts the watermark into a prespecified range of frequencies in the transformed image. The watermark is usually embedded in the perceptually significant portion (which is the significant frequency component) of the image in order to be robust to resist attack. The watermark is scaled according to the magnitude of the particular frequency component. The watermark consists of a random, Gaussian-distributed sequence. This kind of embedding is called *multiplicative watermarking* [11]. If H is the DCT coefficients of the host image and W is the random vector, then $\{h(m,n)\}$ and $\{w(i)\}$ denote their respective pixels. We can embed W into H to become the watermarked image H^* as

$$h^*(m,n) = h(m,n)(1 + \alpha(i) \cdot w(i)). \tag{4.2}$$

Note that a large $\{\alpha(i)\}$ would produce a higher distortion on the watermarked image. It is usually set to be 0.1 to provide a good trade-off between imperceptibility and robustness. An alternative embedding formula using the logarithm of the original coefficients is

$$h^*(m,n) = h(m,n) \cdot e^{\alpha(i) \cdot w(i)}. \tag{4.3}$$

For example, let the watermark be a Gaussian sequence of 1,000 pseudorandom real numbers. We select the 1,000 largest coefficients in the DCT domain, and embed the watermark into a Lena image to obtain the watermarked image, as shown in Figure 4.9. The watermark can be extracted using the inverse embedding formula as

$$w'(i) = \frac{h^*(m,n) - h(m,n)}{\alpha(i) \cdot h(m,n)}. \tag{4.4}$$

For comparison of the extracted watermark sequence $\mathbf{w}'$ and the original watermark $\mathbf{w}$, we can use the similarity measure as

$$sim(\mathbf{w}', \mathbf{w}) = \frac{\mathbf{w}' \cdot \mathbf{w}}{|\mathbf{w}'|}. \tag{4.5}$$

FIGURE 4.9 The watermarked image.

The dot product of $\mathbf{w}' \cdot \mathbf{w}$ will be distributed according to the distribution of a linear combination of variables that are independent and normally distributed as

$$N(0, \mathbf{w}' \cdot \mathbf{w}). \tag{4.6}$$

Thus, $sim(\mathbf{w}', \mathbf{w})$ is distributed according to $N(0,1)$. We can then apply the standard significance tests for the normal distribution. Figure 4.10 shows the response of the watermark detector to 500 randomly generated watermarks, of which only one matches the watermark. The positive response due to the correct watermark is much stronger than the response to incorrect watermarks [11]. If $\mathbf{w}'$ is created independently

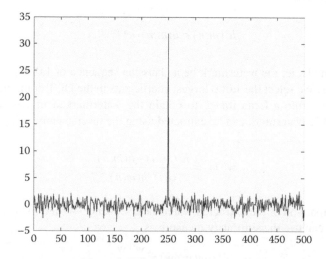

FIGURE 4.10 Responses of 500 randomly generated watermarks, of which only one exactly matches the original watermark.

from **w**, then it is extremely unlikely that $sim(\mathbf{w'},\mathbf{w}) > 6$. This suggests that the embedding technique has very low false-positive and false-negative response rates. By using equation 4.5 on Figure 4.9, we obtain the similarity measure of 29.8520.

4.2.3 WATERMARKING BASED ON VECTOR QUANTIZATION

A comprehensive review on vector quantization (VQ) can be found in Gray [12]. It is a lossy block-based compression technique in which vectors, instead of scalars, are quantized. An image is partitioned into two-dimensional (2D) blocks. Each input block is mapped to a finite set of vectors forming the codebook that is shared by the encoder and the decoder. The index of the vector that best matches the input block based on some cost functions is sent to the decoder. While traditional vector quantization is a fixed-dimension scheme, a more general variable dimension vector quantization has been developed.

A codebook is used to store all the vectors. Given an input vector **v**, Euclidean distance is used in the search process to measure the distance between 2 vectors as

$$k = \arg \min_j \sqrt{\sum_i (\mathbf{v}(i) - \mathbf{s}_j(i))^2}, \quad \text{where } j = 0,1,...,N-1 \quad (4.7)$$

The closest Euclidean distance code word $\mathbf{s}_k$ is selected and transmitted to the recipient. With the same codebook, the decomposition procedure can easily extract the vector **v** via table lookup.

Lu and Sun have presented an image watermarking technique based on VQ [13]. They divided the codebook into a few clusters of close code vectors with the Euclidean distance less than a given threshold. The scheme requires the original image for watermark extraction and uses the codebook partition as the secret key. When the received index is not in the same cluster as the best-match index, a tampering attack is detected.

4.2.4 THE ROUNDING ERROR PROBLEM

The rounding error problem exists in frequency-domain substitution watermarking. A recipient may confront the problem of correctly extracting the embedded watermark because the embedded message is changed due to modification of coefficient values. Ideally, the data after performing the the DCT or DWT, followed by an inverse transformation such as inverse discrete cosine transform (IDCT) or inverse discrete wavelet transform, should be exactly the same as the original data. However, if some data in the transformed image are changed, we will not obtain an image with all integers. This situation is illustrated in Figure 4.11(a), which is the transformed image of Figure 4.7(a). If we apply IDCT to Figure 4.11(a), the original image can be reconstructed as Figure 4.11(b). If the data enclosed by the bold rectangular boxes in Figure 4.11(c) are changed, the pixel values of the reconstructed image will not be integers, as in Figure 4.11(d). Note that the changes between Figures 4.11(a) and Figure 4.11(c) are small.

Some researchers suggest that the rounding technique be used to convert the real numbers to integers. Therefore, after adopting the rounding approach to Figure 4.12(a), we show the rounded image in Figure 4.12(b). If the recipient wants

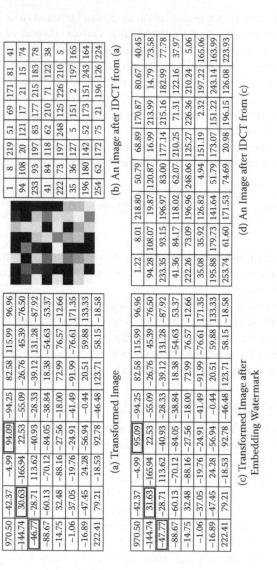

(a) Transformed Image

970.50	−42.37	−4.99	94.09	−94.25	82.58	115.99	96.96
−144.74	30.63	−165.94	22.53	−55.09	−26.76	45.39	−76.50
−46.77	−28.71	113.62	−40.93	−28.33	−39.12	131.28	−87.92
−88.67	−60.13	−70.12	−84.05	−38.84	18.38	−54.63	53.37
−14.75	32.48	−88.16	−27.56	−18.00	72.99	76.57	−12.66
−1.06	−37.05	−19.76	−24.91	−41.49	−91.99	−76.61	171.35
−16.89	−47.45	24.28	−56.94	−0.44	20.51	59.88	133.33
222.41	79.21	−18.53	92.78	−46.48	123.71	58.15	−18.58

(b) An Image after IDCT from (a)

1	8	219	51	69	171	81	41
94	108	20	121	17	21	15	74
233	93	197	83	177	215	183	78
41	84	118	62	210	71	122	38
222	73	197	248	125	226	210	5
35	36	127	5	151	2	197	165
196	180	142	52	173	151	243	164
254	62	172	75	21	196	126	224

(c) Transformed Image after Embedding Watermark

970.50	−42.37	−4.99	95.09	−94.25	82.58	115.99	96.96
−144.74	31.63	−165.94	22.53	−55.09	−26.76	45.39	−76.50
−47.77	−28.71	113.62	−40.93	−28.33	−39.12	131.28	−87.92
−88.67	−60.13	−70.12	−84.05	−38.84	18.38	−54.63	53.37
−14.75	32.48	−88.16	−27.56	−18.00	72.99	76.57	−12.66
−1.06	−37.05	−19.76	−24.91	−41.49	−91.99	−76.61	171.35
−16.89	−47.45	24.28	−56.94	−0.44	20.51	59.88	133.33
222.41	79.21	−18.53	92.78	−46.48	123.71	58.15	−18.58

(d) An Image after IDCT from (c)

1.22	8.01	218.80	50.79	68.89	170.87	80.67	40.45
94.28	108.07	19.87	120.87	16.99	213.99	14.79	73.58
233.35	93.15	196.97	83.00	177.14	215.16	182.99	77.78
41.36	84.17	118.02	62.07	210.25	71.31	122.16	37.97
222.26	73.09	196.96	248.06	125.27	226.36	210.24	5.06
35.08	35.92	126.82	4.94	151.19	2.32	197.22	165.06
195.88	179.73	141.64	51.79	173.07	151.22	243.14	163.99
253.74	61.60	171.53	74.69	20.98	196.15	126.08	223.93

FIGURE 4.11 An illustration that the pixel values in the reconstructed image will not be integers if some coefficient values are changed.

1.22	8.01	218.80	50.79	68.89	170.87	80.67	40.45
94.28	108.07	19.87	120.87	16.99	213.99	14.79	73.58
233.35	93.15	196.97	83.00	177.14	215.16	182.99	77.78
41.36	84.17	118.02	62.07	210.25	71.31	122.16	37.97
222.26	73.09	196.96	248.06	125.27	226.36	210.24	5.06
35.08	35.92	126.82	4.94	151.19	2.32	197.22	165.06
195.88	179.73	141.64	51.79	173.07	151.22	243.14	163.99
253.74	61.60	171.53	74.69	20.98	196.15	126.08	223.93

(a) Image after IDCT Transform

1	8	219	51	69	171	81	40
94	108	20	121	17	214	15	74
233	93	197	83	177	215	183	78
41	84	118	62	210	71	122	38
222	73	197	248	125	226	210	5
35	36	127	5	151	2	197	165
196	180	142	52	173	151	243	164
254	62	172	75	21	196	126	224

(b) Translate Real Numbers into Integers by ROUND

970.38	-42.20	-5.16	94.24	-94.38	82.68	115.92	96.99
-144.91	30.87	-166.17	22.73	-55.26	-26.62	45.30	-76.46
-46.94	-28.49	113.41	-40.74	-28.49	-38.99	131.19	-87.88
-88.82	-59.93	-70.31	-83.88	-38.99	18.49	-54.71	53.41
-14.88	32.65	-88.32	-27.41	-18.13	73.09	76.50	-12.62
-1.16	-36.91	-19.89	-24.79	-41.58	-91.91	-76.66	171.37
-16.95	-47.36	24.19	-56.86	-0.51	20.56	59.84	133.35
222.37	79.26	-18.58	92.82	-46.51	123.73	58.13	-18.57

(c) Transform Rounded Image by DCT

-	1	0	-	-
-	0	-	-	-
0	-	Original Watermark 1101	-	-
-	-	-	-	-
-	-	-	-	-

(d) Extract the Embedded Watermark

FIGURE 4.12 An example showing that the watermark cannot be correctly extracted.

TABLE 4.1
Total Possible Embedded Watermarks

Embedded	Extracted	Error Bits	Embedded	Extracted	Error Bits
0000	0000	0	1000	0000	1
0001	0000	1	1001	0000	2
0010	0000	1	1010	0000	2
0011	0000	2	1011	0000	3
0100	0000	1	1100	0000	2
0101	0000	2	1101	0010	4
0110	0000	2	1110	1110	0
0111	1110	2	1111	1110	1

to extract the watermark from the rounded image, she will perform a DCT to the rounded image and obtain its transformed image, as shown in Figure 4.12(c). Unfortunately, the recipient cannot correctly extract the watermark from the location where it is embedded even though there is only one difference, enclosed by the bold rectangle, between Figure 4.12(b) and its original image, Figure 4.7(a).

Table 4.1 shows the results of embedding different kinds of watermarks into the same transformed image in Figure 4.7(d). The embedding rule is shown in Figure 4.13. For a 4-bit watermark 1234, we insert the most significant bit 1 into position A, 2 into position B, 3 into position C, and 4 into position D. There are 2^4 possible embedding combinations. Among the 16 possibilities, only two cases of watermarks can be extracted correctly.

4.3 THE FRAGILE WATERMARK

Due to the popularity in computerized image processing, there are a tremendous number of software tools available for users to employ in modifying images. Therefore, it becomes a critical problem for a recipient to judge whether the received image is altered by an attacker. This situation is illustrated in Figure 4.14.

-	-	C	D	-	-	-	-
-	B	-	-	-	-	-	-
A	-	-	-	-	-	-	-
-	-	-	-	-	-	-	-
-	-	-	-	-	-	-	-
-	-	-	-	-	-	-	-
-	-	-	-	-	-	-	-
-	-	-	-	-	-	-	-

FIGURE 4.13 The positions where the watermark is embedded.

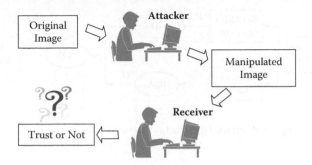

FIGURE 4.14 A recipient has a problem of judging whether a received image is altered.

The fragile watermark provides a solution for ensuring if a received image can be trusted. The recipient will evaluate the hidden fragile watermark to see whether the image is altered. Therefore, the fragile watermarking scheme is used to detect any unauthorized modification.

4.3.1 The Block-Based Fragile Watermark

Wong has presented a block-based fragile watermarking algorithm that is capable of detecting changes [14], such as pixel values and image size, by adopting the Rivest–Shamir–Adleman (RSA) public-key encryption algorithm and Message Digest 5 (MD5) for the hash function [15–16]. Wong's watermarking insertion and extraction algorithms are introduced below, and their flowcharts are shown in Figures 4.15 and 4.16, respectively.

Wong's Watermarking Insertion Algorithm

1. Divide the original image X into subimages X_r.
2. Divide the watermark W into subwatermarks W_r.
3. For each subimage X_r we obtain X'_r by setting the LSB to be 0.
4. For each X'_r we obtain the corresponding codes H_r by a cryptographic hash function (e.g., MD5).
5. XOR_r is obtained by adopting the *exclusive or* (XOR) operation of H_r and W_r.

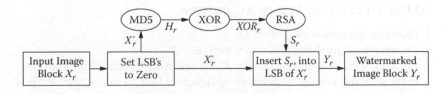

FIGURE 4.15 Wong's watermarking insertion algorithm.

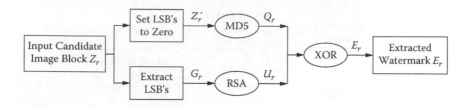

FIGURE 4.16 Wong's watermarking extraction algorithm.

6. S_r is obtained by encrypting XOR_r using the RSA encryption with a private key K'.
7. We obtain the watermarked subimage Y_r by embedding S_r into LSB of X'_r.

WONG'S WATERMARKING EXTRACTION ALGORITHM

1. Divide the candidate image Z into subimages Z_r.
2. Z'_r is obtained by setting LSB of Z_r to be 0.
3. G_r is obtained by extracting LSB of Z_r.
4. U_r is obtained by decrypting G_r using RSA with a public key K.
5. For each Z'_r we obtain the corresponding codes Q_r by a cryptographic hash function (e.g., MD5 (Message-Digest algorithm 5) or SHA (Secure Hash Algorithm)).
6. E_r is the extracted watermark by adopting the XOR operation of Q_r and G_r.

4.3.2 WEAKNESSES OF THE BLOCK-BASED FRAGILE WATERMARK

Holliman and Memon have developed a VQ counterfeiting attack to forge the fragile watermark by exploiting the blockwise independence of Wong's algorithm [17]. Through this exploitation we can rearrange blocks of an image to form a new collage image in which the embedded fragile watermark is not altered. Therefore, given a large database of watermarked images, the VQ attack can approximate a counterfeit collage image with the same visual appearance as an original unwatermarked image from a codebook. The VQ attack does not need to have the knowledge of the embedded watermarks.

If D is the large database of watermarked images and C the codebook with n items generated from D, the algorithm of the VQ attack can be described as follows (its flowchart is shown in Figure 4.17):

VQ COUNTERFEITING ATTACK ALGORITHM

1. Generate the codebook C from D.
2. Divide the original image X into subimages X_r.
3. For each subimage X_r, we obtain X_r^c by selecting the data from the codebook with the minimum distance (difference) to X_r. Finally, we obtain a counterfeit collage image X^c with the same visual appearance as the original unwatermarked image.

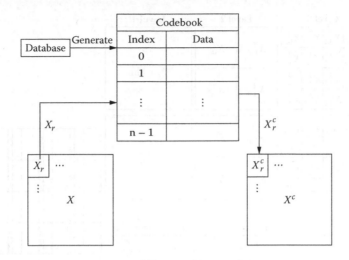

FIGURE 4.17 The VQ counterfeit attack.

4.3.3 THE HIERARCHICAL BLOCK-BASED FRAGILE WATERMARK

In order to defeat the VQ counterfeiting attack, a hierarchical block-based watermarking technique was developed by Celik et al. [18]. The idea of the hierarchical watermarking approach is simply to break the blockwise independence of Wong's algorithm. That is, the embedded data not only include the information of the corresponding block, but also possess the relative information of the higher-level blocks. Therefore, the VQ attack cannot approximate an image based on the codebook that only records the information of each watermarked block.

Let $X_{i,j}^l$ denote a block in the hierarchical approach, where (i, j) represents the spatial position of the block and l is the level to which the block belongs. The total number of levels in the hierarchical approach is denoted by L. At each successive level, the higher-level block is divided into 2×2 lower-level blocks. That is, for $l = L\text{-}1$ to 2,

$$\begin{bmatrix} X_{2i,\,2j}^{l+1} & X_{2i,\,2j+1}^{l+1} \\ X_{2i+1,\,2j}^{l+1} & X_{2i+1,\,2j+1}^{l+1} \end{bmatrix} = X_{i,j}^l.$$

For each block $X_{i,j}^l$, we obtain its corresponding *ready-to-insert data* (RID), $S_{i,j}^l$, after the processes of MD5, XOR, and RSA. Then we construct a payload block $P_{i,j}^L$ based on the lowest-level block $X_{i,j}^L$. For each payload block, it contains both RID and the data belonging to the higher-level block of $X_{i,j}^L$.

Figure 4.18 illustrates an example of the hierarchical approach; Figure 4.18(a) denotes an original image X and its three-level hierarchical results. Note that, $X_{0,0}^1$ is the top level of the hierarchy, consisting of only one block, X. Figure 4.18(b) shows the corresponding RID of each block $X_{i,j}^l$. In Figure 4.18(c), when dealing with the block $X_{3,3}^3$, we consider the following three RIDs: $S_{3,3}^3$, $S_{1,1}^2$, and $S_{0,0}^1$. After generating the payload block $P_{3,3}^3$, we embed it into the LSB of the original image X on $X_{3,3}^3$.

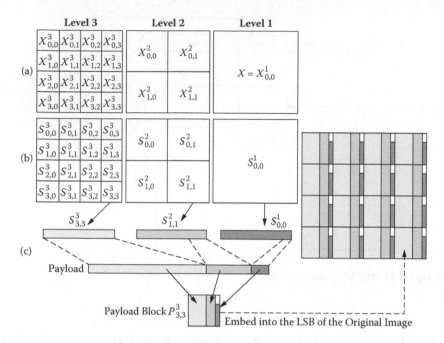

FIGURE 4.18 An example of the hierarchical approach with three levels.

4.4 THE ROBUST WATERMARK

Computer technology allows people to easily duplicate and distribute digital multimedia through the Internet. However, these benefits bring concomitant risks of data piracy. One solution to providing security for copyright protection is robust watermarking, which ensures that embedded messages survive some attackers, such as JPEG compression, Gaussian noise, and low-pass filtering. Figure 4.19 illustrates the purpose of robust watermarking.

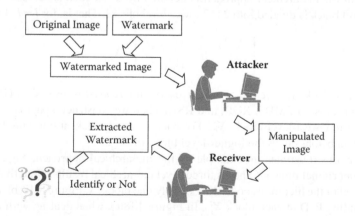

FIGURE 4.19 The purpose of robust watermarking.

215	201	177	145	111	79	55	41
201	177	145	111	79	55	41	215
177	145	111	79	55	41	215	201
145	111	79	55	41	215	201	177
111	79	55	41	215	201	177	145
79	55	41	215	201	177	145	111
55	41	215	201	177	145	111	79
41	215	201	177	145	111	79	55

1024.0	0.0	0.0	0.0	0.0	0.0	0.0	0.0
0.0	110.8	244.0	-10.0	35.3	-2.6	10.5	-0.5
0.0	244.0	-116.7	-138.2	0.0	-27.5	0.0	-6.4
0.0	-10.0	-138.2	99.0	72.2	0.2	13.3	0.3
0.0	35.3	0.0	72.2	-94.0	-48.2	0.0	-7.0
0.0	-2.6	-27.5	0.2	-48.2	92.3	29.9	0.4
0.0	10.5	0.0	13.3	0.0	29.9	-91.3	-15.7
0.0	-0.5	-6.4	0.3	-7.0	0.4	-15.7	91.9

FIGURE 4.20 An example of significant coefficients.

4.4.1 THE REDUNDANT EMBEDDING APPROACH

It is obvious that the embedded message will be distorted due to some image processing procedures from attackers. In order to achieve robustness, we embed redundant watermarks in a host image [19]. Epstein and McDermott hold a patent for copy protection via redundant watermark encoding [20].

4.4.2 THE SPREAD SPECTRUM APPROACH

Cox et al. have presented a spread spectrum–based robust watermarking algorithm in which the embedded messages are spread throughout the image [21, 22]. Their algorithm first translates an image into its frequency domain; then the watermarks are embedded into significant coefficients of the transformed image. The significant coefficients are the locations of large absolute values of the transformed image, as shown in Figure 4.20. Figure 4.20(a) shows an original image, and 4.20(b) is its image transformed by DCT. The threshold is 70. The bold rectangles in Figure 4.20(b) are called *significant coefficients* since the absolute values are larger than 70.

REFERENCES

[1] Podilchuk, C. I., and E. J. Delp. "Digital Watermarking: Algorithms and Applications." *IEEE Signal Processing Mag.* 18 (2001): 33.

[2] Pitas, I. A Method for Watermark Casting on Digital Images." *IEEE Trans. Circuits and Systems for Video Technology* 8 (1998): 775.

[3] Wolfgang, R., and E. Delp. "A Watermarking Technique for Digital Imagery: Further Studies." In *Proc. Int. Conf. Imaging Science, Systems and Technology*. Las Vegas, NV, 1997.

[4] Shih, F. Y., and S. Y. Wu. "Combinational Image Watermarking in the Spatial and Frequency Domains." *Pattern Recognition* 36 (2003): 969.

[5] Wong, P. W. "A Public Key Watermark for Image Verification and Authentication." In *Proc. IEEE Int. Conf. Image Processing*. Chicago, 1998.

[6] Wolfgang, R., and E. Delp. "A Watermark for Digital Images." In *Proc. IEEE Intl. Conf. Image Processing*. Lausanne, Switzerland: IEEE, 1996.

[7] Lin, E., and E. Delp, "Spatial Synchronization Using Watermark Key Structure." In *Proc. SPIE Conf. Security, Steganography, and Watermarking of Multimedia Contents*. San Jose, CA: SPIE, 2004.

[8] Mukherjee, D., S. Maitra, and S. Acton. "Spatial Domain Digital Watermarking of Multimedia Objects for Buyer Authentication." *IEEE Trans. on Multimedia* 6 (2004): 1.

[9]   Lin, S. D., and C.-F. Chen. "A Robust DCT-Based Watermarking for Copyright Protection." *IEEE Trans. Consumer Electronics* 46 (2000): 415.

[10] Huang, J., Y. Shi, Y., and Y. Shi. Embedding Image Watermarks in DC Components. *IEEE Trans. Circuits and Systems for Video Technology* 10 (2000): 974.

[11] Cox, J., et al. "A Secure, Robust Watermark for Multimedia." In *Proc. First Int. Workshop Information Hiding*. Cambridge, UK, 1996, 185.

[12] Gray, R. M. "Vector Quantization." *IEEE ASSP Mag.* 1 (1984): 4.

[13] Lu, Z., and S. Sun. "Digital Image Watermarking Technique Based on Vector Quantization." *IEEE Electronics Letters* 36 (2000): 303.

[14] Wong, P. W. "A Public Key Watermark for Image Verification and Authentication." In *Proc. IEEE Int. Conf. Image Processing*. Chicago, 1998.

[15] Rivest, R., A. Shamir, and L. Adleman. "A Method for Obtaining Digital Signatures and Public-Key Cryptosystems." *Communications of the ACM* 21 (1978): 120.

[16] Rivest, R. L. *The MD5 Message Digest Algorithm*. RFC 1321. Cambridge, MA: MIT Laboratory for Computer Science/RSA Data Security 1992.

[17] Holliman, M., and N. Memon. Counterfeiting Attacks on Oblivious Block-Wise Independent Invisible Watermarking Schemes. *IEEE Trans. Image Processing* 9 (2000): 432.

[18] Celik, M., et al. "Hierarchical Watermarking for Secure Image Authentication with Localization." *IEEE Trans. Image Processing* 11 (2002): 585.

[19] Coetzee, L., and J. Eksteen. "Copyright Protection for Cultureware Preservation in Digital Repositories." In *Proc. 10th Annual Internet Society Conference*. Yokohama, Japan, 2000.

[20] Epstein, M., and R. McDermott. "Copy Protection via Redundant Watermark Encoding. U.S. Patent no. 7133534, 2006. Available online at http://www.patentgenius.com/patent/7133534.html.

[21] Cox, I. J., et al. "Secure Spread Spectrum Watermarking for Multimedia." *IEEE Trans. Image Processing* 6 (1997): 1673.

[22] Cox, I. J., et al. "Digital Watermarking." *J. Electronic Imaging* 11 (2002): 414.

5 Watermarking Attacks and Tools

There are a great number of digital watermarking methods developed in recent years that embed secret watermarks in images for various applications. The purpose of watermarking can be robust or fragile, private or public. A *robust* watermark aims to overcome the attacks from common image processing or intentional attempts to remove or alter it [1–3]. A *fragile* watermark aims to sense whether the watermarked image has been altered—even slightly. A *private* watermark is used when the original unmarked image is available for comparison. However, a *public* watermark is set up in an environment in which the original unmarked image is unavailable for watermark detection.

The attackers aiming at watermarked images can be classified as unintentional or intentional. An attack action can disable a user's watermarking target in a way that the embedded message cannot function well. The unintentional attacker may not be aware of affecting the original watermark; for example, one could reduce an image's size through JPEG compression or may choose to smooth out the image a bit. The intentional attacker malignantly disables the function of watermarks for a specific reason; for example, one could alter or destroy the hidden message for an illegal purpose such as image forgery.

In order to identify weaknesses in watermarking techniques, propose improvements, and study the effects of current technology on watermarking, we need to understand the varieties of attacks on watermarked images. This chapter will discuss different types of watermarking attacks and available tools. Since there are many different types of attacks for a variety of purposes, not everything will be covered here. But examples will be given of some popular attack schemes. We can categorize these various watermarking attack schemes into four classes:

1. image processing attacks
2. geometric transformation
3. cryptographic attacks
4. protocol attacks

5.1 IMAGE PROCESSING ATTACKS

Image processing attacks include filtering, remodulation, JPEG coding distortion, and JPEG 2000 compression.

5.1.1 ATTACKS BY FILTERING

Filtering indicates the image processing operation in frequency domain. We first compute the Fourier transform of the image, multiply the result by a filter transfer function, and then perform the inverse Fourier transform. This idea behind *sharpening*

FIGURE 5.1 (a) The watermarked image. (b) After the 3 × 3 sharpening filter attack.

is to increase the magnitude of high-frequency components relative to low-frequency components. The high-pass filter in the frequency domain is equivalent to the impulse shape in the spatial domain. Therefore, the sharpening filter in the spatial domain contains positive values close to the center and negative values surrounding the outer boundary. A typical 3 × 3 sharpening filter is

$$\frac{1}{9}\begin{bmatrix} -1 & -1 & -1 \\ -1 & 8 & -1 \\ -1 & -1 & -1 \end{bmatrix}. \tag{5.1}$$

An example of 3 × 3 sharpening filter attack is shown in Figure 5.1.

The idea behind *blurring* is to decrease the magnitude of high-frequency components. The low-pass filter in the frequency domain is equivalent to the mountain shape in the spatial domain. Therefore, the smoothing filter in the spatial domain should have all positive coefficients, with the largest at the center. The simplest low-pass filter would be a mask with all coefficients having a value of 1. A sampled 3 × 3 Gaussian filter is

$$\frac{1}{12}\begin{bmatrix} 1 & 1 & 1 \\ 1 & 4 & 1 \\ 1 & 1 & 1 \end{bmatrix}. \tag{5.2}$$

A sampled 5 × 5 Gaussian filter is

$$\frac{1}{273}\begin{bmatrix} 1 & 4 & 7 & 4 & 1 \\ 4 & 16 & 26 & 16 & 4 \\ 7 & 26 & 41 & 26 & 7 \\ 4 & 16 & 26 & 16 & 4 \\ 1 & 4 & 7 & 4 & 1 \end{bmatrix}. \tag{5.3}$$

(a)　　　　　　　　　　　　　　　　　(b)

FIGURE 5.2 (a) The watermarked image. (b) After the 5 × 5 Gaussian filter attack.

An example of conducting the 5 × 5 Gaussian filter attack is shown in Figure 5.2.

In image enhancement, a median filter is often used to achieve noise reduction rather than blurring. The method is rather simple: we calculate the median value of the gray values surrounding the pixel's vicinity and assign it to the pixel. An example of using a 5 × 5 median filtering is shown in Figure 5.3.

5.1.2 Attack by Remodulation

A remodulation attack was first presented by Langelaar et al. [4]. In this scheme the watermark is predicted via subtraction of the median filtered version of the watermarked image. The predicted watermark was additionally high-pass filtered, truncated, and then subtracted from the watermarked image with a constant amplication factor of 2. The basic remodulation attack consists of removing noise from the image, inverting the estimated watermark, and adding it to a percentage of certain locations. If the watermark is additive, this attack is quite effective and leaves few artifacts. In most cases the attacked image contains little distortion. An example of remodulation attack is shown in Figure 5.4.

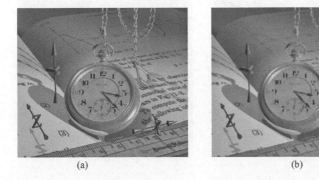

(a)　　　　　　　　　　　　　　　　(b)

FIGURE 5.3 (a) The watermarked image. (b) After the 5 × 5 median filtering attack.

(a)   (b)

FIGURE 5.4 (a) The watermarked image. (b) After the 3 × 3 remodulation attack. (From Pereira, S. et al. in *Proc. Information Hiding Workshop III*, Pittsburgh, PA. 2001.)

5.1.3 ATTACK BY JPEG CODING DISTORTION

JPEG—which stands for the Joint Photographic Experts Group, which created the technology—is very popular in color-image compression. It was created in 1986, issued in 1992, and approved by the International Organization for Standardization in 1994. Often used in Internet browsers is the progressive JPEG, which reorders information in such a way that an obscure perception of the whole image is available rather than a brittle perception of just a small portion after only a small part of the image has been downloaded. Although JPEG could reduce an image file size to about 5% of its normal size, some detail may be lost in the compressed image. Figure 5.5 shows a JPEG-encoded version of a Lena image with a quality of 10.

5.1.4 ATTACK BY JPEG 2000 COMPRESSION

JPEG 2000, which is a standard image compression technique created by the JPEG committee, uses state-of-the-art wavelet-based technology. This creation aims at overcoming the blocky appearance problem occurred in the discrete cosine transform–based JPEG standard. Its architecture provides various applications

(a)   (b)

FIGURE 5.5 (a) The watermarked image. (b) After the JPEG attack.

(a)   (b)

FIGURE 5.6   (a) The watermarked image. (b) After the JPEG 2000 attack.

ranging from prepublishing, digital photography, medical imaging, and other industrial imaging. In general, JPEG 2000 is able to achieve a high compression ratio while simultaneously avoiding blocky and blurry artifacts. It also allows more sophisticated progressive downloads. An example of the JPEG 2000 attack with 0.5 bits per pixel is shown in Figure 5.6.

5.2   GEOMETRIC TRANSFORMATION

Geometric attacks include scaling, rotation, clipping, linear transformation, bending, warping, perspective projection, collage, and template. O'Ruanaidh and Pun have proposed digital watermarking technology using Fourier–Mellin transform-based invariants [5], which do not need the original image to extract the embedded watermark. According to their design, the embedded watermark will not be destroyed by the combined rotation, scale, and translation transformations. The original image is not required for extracting the embedded mark.

5.2.1   ATTACK BY IMAGE SCALING

Sometimes when we scan a printed image or adjust its size for electronic publishing, image scaling may occur. This should be especially noted as we move increasingly in the direction of web publishing. An example is when the watermarked image is first scaled down (or down-sampled) by reducing both length and width by one-half—that is, by averaging every 2×2 block into a pixel. This quarter-sized image is then scaled up (or up-sampled) to its original size through the interpolation method. In general, bilinear interpolation will yield a smooth appearance, and bicubic interpolation will yield even smoother results. An example of the scaling attack is shown in Figure 5.7.

Image scaling can be of one of two types: uniform or nonuniform. Uniform scaling uses the same scaling factors in horizontal and vertical directions. However, nonuniform scaling applies different scaling factors in horizontal and vertical directions. In other words, the aspect ratio is changed in nonuniform scaling. An example of nonuniform scaling using xscale = 0.8 and yscale = 1 is shown in Figure 5.8. Note that the nonuniform scaling will produce the effect of object shape distortion. Most digital watermarking methods are designed to be flexible only to uniform scaling.

(a)   (b)

FIGURE 5.7 (a) The watermarked image. (b) After the scaling attack.

5.2.2 ATTACK BY ROTATION

Suppose that we want to apply a rotation angle θ to an image $f(x, y)$. Let the counterclockwise rotation angle be positive and the image after rotation be $f^*(x^*, y^*)$. Figure 5.9 illustrates the rotation from a point P to P^*. If R denotes the distance from P to the origin, we can derive the following formulae:

$$x = r\cos\phi \qquad y = r\sin\phi$$

$$x^* = r\cos(\theta + \phi) = r\cos\phi\cos\theta - r\sin\phi\sin\theta$$

$$= x\cos\theta - y\sin\theta$$

$$y^* = r\sin(\theta + \phi) = r\cos\phi\sin\theta + r\sin\phi\cos\theta$$

$$= x\sin\theta + y\cos\theta$$

$$[x^* \; y^*] = [x \; y]\begin{bmatrix} \cos\theta & \sin\theta \\ -\sin\theta & \cos\theta \end{bmatrix}$$

(a)   (b)

FIGURE 5.8 (a) The watermarked image. (b) After the nonuniform scaling attack. (From Pereira, S. et al. in *Proc. Information Hiding Workshop III*, Pittsburgh, PA. 2001.)

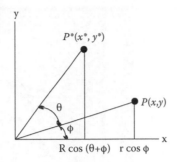

FIGURE 5.9   An illustration of the rotation.

When we perform the formulae to compute the new coordinates, they need to be converted into integers. A simple rounding approach will often produce unacceptable results, since some points of the new image may have no gray levels assigned to them by this method while others may have more than one gray level assigned. To solve these problems, the inverse rotation and interpolation methods are needed (for details, see [6]).

When we apply a small angle rotation, often combined with cropping, on a watermarked image, the overall perceived content of the image is the same but the embedded watermark may be undetectable. When a print image is scanned, it is often necessary to use a rotation to adjust a slight tilt. An example of applying a 10° counterclockwise rotation to the watermarked image is shown in Figure 5.10.

5.2.3   ATTACK BY IMAGE CLIPPING

Figure 5.11(a) shows a clipped version of the watermarked image. Only the central quarter of the image remains. In order to extract the watermark from this image, the missing portions of the image are replaced by the corresponding portions from the original unwatermarked image, as shown in Figure 5.11(b).

(a)   (b)

FIGURE 5.10   (a) The watermarked image. (b) After the rotational attack. (From Pereira, S. et al. in *Proc. Information Hiding Workshop III*, Pittsburgh, PA. 2001.)

(a)   (b)

FIGURE 5.11   (a) A clipped version of a JPEG-encoded (10% quality) Lena image. (b) The restored version.

5.2.4   ATTACK BY LINEAR TRANSFORMATION

We can apply a 2×2 linear transformation matrix. Let x and y denote the original coordinates, and x^* and y^* denote the transformed coordinates. The transformation matrix T to change (x, y) into (x^*, y^*) can be expressed as

$$[x^* \, y^*] = [x \, y].\begin{bmatrix} T_{11} & T_{12} \\ T_{21} & T_{22} \end{bmatrix}. \tag{5.4}$$

An example of using $T_{11} = 1.15$, $T_{12} = -0.02$, $T_{21} = -0.03$, and $T_{22} = 0.9$ is shown in Figure 5.12.

(a)   (b)

FIGURE 5.12   (a) The watermarked image. (b) After the linear transformation attack. (From Pereira, S. et al. in *Proc. Information Hiding Workshop III*, Pittsburgh, PA. 2001.)

(a)   (b)

FIGURE 5.13 (a) The watermarked image. (b) After the bending attack. (From Pereira, S. et al. in *Proc. Information Hiding Workshop III*, Pittsburgh, PA. 2001.)

5.2.5 ATTACK BY BENDING

This attack was originally proposed by Petitcolas [7]. Local nonlinear geometric distortions are applied and the image is interpolated. Noise is also added and a small compression is applied. An example of the bending attack is shown in Figure 5.13.

5.2.6 ATTACK BY WARPING

Image warping performs a pixel-by-pixel remapping (or "warp") of an input image to an output image. It requires an integer-sampled "remapping image" that is the same size as the output image. This remapping image consists of x- and y-channels. For each pixel in the output image, the corresponding point in the remapping image is checked to determine which pixel from the input image will be copied. The image grid is first warped into three dimensions and then projected back onto two dimensions. An example of the warping attack with a factor of 3 is shown in Figure 5.14.

5.2.7 ATTACK BY PERSPECTIVE PROJECTION

In contrast to parallel projection, in perspective projection the parallel lines converge, object size is reduced with increasing distance from the center of projection, and nonuniform foreshortening of lines in the object as a function of orientation and

(a)   (b)

FIGURE 5.14 (a) The watermarked image. (b) After the warping attack. (From Pereira, S. et al. in *Proc. Information Hiding Workshop III*, Pittsburgh, PA. 2001.)

(a)

(b)

FIGURE 5.15 (a) The watermarked image. (b) After the perspective projection attack. (From Pereira, S. et al. in *Proc. Information Hiding Workshop III*, Pittsburgh, PA. 2001.)

distance of the object from the center of projection occurs. All of these effects aid the depth perception of the human visual system, but the object shape is not preserved. An example of this attack using 30° rotation along an x-axis together with a perspective projection is shown in Figure 5.15.

5.2.8 ATTACK BY COLLAGE

The collage attack as proposed by Holliman and Memon on blockwise independent watermarking techniques [8] assumed that the watermark logo would be available to the attacker. They used multiple authenticated images to develop a new image by combining parts of different images with a preservation of their relative spatial locations in the image with each block being embedded independently. An example of the collage attack is shown in Figure 5.16.

5.2.9 ATTACK BY TEMPLATE

Templates are used as a synchronization pattern to estimate the applied affine transformation [9]. We then apply the inverse transformation to extract the watermark

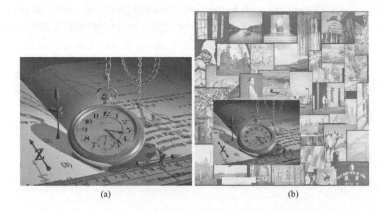
(a)   (b)

FIGURE 5.16 (a) The watermarked image. (b) After the collage attack. (From Pereira, S. et al. in *Proc. Information Hiding Workshop III*, Pittsburgh, PA. 2001.)

based on the obtained estimation. Many modern schemes use a template in the frequency domain to extract the watermark. The template is composed of peaks, and the watermark detection process consists of two steps. First, the affine transformation undergone by the watermarked image is calculated. Second, the inverse transformation is applied to decode the watermark.

5.3 CRYPTOGRAPHIC ATTACK

The principle of cryptographic attack is similar to the attack applied in cryptography, but with different purposes. This form of attack attempts to circumvent the security of a cryptographic system by retrieving the key or exposing the plaintext in a code, cipher, or protocol. Based on exhaustive key-search approaches, the cryptographic attack tries to find out the key used for embedding. When the key is found, the watermark is overwritten. Another cryptographic attack, called *Oracle attack*, tries to create a nonwatermarked image when a watermark detector device is available. The attacked data set is generated by tacking a small portion of each data set and establishing a new data set for attacking. In other words, an attacker can remove a watermark if a public decoder is available by applying small changes to the image until the decoder cannot find it anymore. Linnartz and Dijk have presented analysis of the sensitivity attack against electronic watermarks in images [10].

Another cryptographic attack, called a *collusion attack*, is used when instances of the same data are available. An example of attack by collusion is to take two separately watermarked images and average them to form another, as seen in Figure 5.17.

5.4 PROTOCOL ATTACKS

Protocol attacks intend to generate protocol ambiguity in the watermarking process. An example is based on the idea of invertible watermarks to subtract a designed watermark from the watermarked image and then declare it as its owner. This can cause ambiguity regarding true ownership. One protocol attack, called *copy attack* [11], intends to estimate a watermark from watermarked data and copy it to

(a) (b)

FIGURE 5.17 (a) The watermarked image. (b) After the collusion attack.

some other data instead of destroying the watermark or impairing its detection. The copy attack consists of four stages: (1) estimation of the watermark from the original image; (2) estimation of the mask from the target image; (3) reweighting of the watermark as a function of the mask; and (4) addition of the reweighted watermark onto the target image.

5.5   WATERMARKING TOOLS

This section briefly describes four benchmarking tools for image watermarking: Stirmark, Checkmark, Optimark and Certimark.

Stirmark was developed in November 1997 as a generic tool for evaluating the robustness of image watermarking algorithms. It applies random bilinear geometric distortions in order to desynchronize watermarking algorithms. After that, enhanced versions have been developed to improve the original attack. Stirmark can generate a set of altered images based on the watermarked to identify whether the embedded watermark is detectable [7,12–14]. Stirmark uses a combination of different detection results to determine an overall score. There are nine types of attacks available in Stirmark: compression, cropping, scaling, shearing, rotation, signal enhancement, linear transformation, random geometric distortion, and other geometric transformations.

Checkmark [15], which can run on Matlab under Unix and Windows, provides a benchmarking tool to evaluate the performance of image watermarking techniques. It includes some additional attacks that are unavailable in Stirmark. It has a characteristic of considering different applications in image watermarking and placing different weights based on their importance for a particular usage.

Optimark [16], which was developed at Aristotle University in Thessaloniki, Greece, can evaluate the performance of still-image watermarking algorithms. Its main features include multiple trials of different messages and keys, evaluations of detection/decoding metrics, a graphical user interface, an algorithm breakdown evaluation, and a resulting summarization of selected attacks.

Certimark was developed by a European project, including 15 academic and industrial partners, in 2000. It intends to perform the international certification process on watermarking algorithms as the benchmark reference, and is used as a basis for accessing control of a given watermarking technique. It is believed that the clear framework in the technological assessment will lead to competition between technology suppliers and in the meantime will warrant the robustness of watermarking as measured by the benchmark.

REFERENCES

[1] Cox, J., et al. "A Secure, Robust Watermark for Multimedia." In *Proc. IEEE International Conference on Image Processing*. Lausanne, Switzerland: IEEE, 1996.

[2] Kundur, D., and D. Hatzinakos. "Diversity and Attack Characterization for Improved Robust Watermarking." *IEEE Trans. Signal Processing* 29 (2001): 2383.

[3] Fei, C., D. Kundur, and R. H. Kwong. "Analysis and Design of Secure Watermark-Based Authentication Systems." *IEEE Trans. Information Forensics and Security* 1 (2006): 43.

[4]   Langelaar, G. C., R. L. Lagendijk, and J. Biemond. "Removing Spatial Spread Spectrum Watermarks by Non-Linear Filtering." In *Proc. Europ. Conf. Signal Processing*. Rhodes, Greece, 1998.

[5]   O'Ruanaidh, J., and T. Pun. Rotation, Translation and Scale Invariant Digital Image Watermarking." In *Proc. Int. Conf. Image Processing*. 1997, 536.

[6]   Rosenfeld, A., and A. C. Kak. *Digital Picture Processing*. New York: Academic Press, 1982.

[7]   Petitcolas, F., R. Anderson, and M. Kuhn. "Attacks on Copyright Marking Systems." In *Proc. Int. Workshop on Information Hiding*. Portland, OR, 1998.

[8]   Holliman, M., and N. Memon. "Counterfeiting Attacks for Block-Wise Independent Watermarking Techniques." *IEEE Trans. Image Processing* 9 (2000): 432.

[9]   Pereira, S., et al. "Template-Based Recovery of Fourier-Based Watermarks Using Log-Polar and Log-Log Maps." In *Proc. IEEE Int. Conf. Multimedia Computing and Systems*. Florence, Italy, 1999.

[10] Linnartz, J., and M. Dijk, M. Analysis of the Sensitivity Attack against Electronic Watermarks in Images." In *Proc. Intl. Workshop on Information Hiding*. Portland, OR, 1998.

[11] Deguillaume, F., et al. "Secure Hybrid Robust Watermarking Resistant against Tampering and Copy Attack." *Signal Processing* 83 (2003): 2133.

[12] Kutter, M., and F. Petitcolas. "A Fair Benchmark for Image Watermarking Systems." In *Proc. SPIE Electronic Imaging 1999: Security and Watermarking of Multimedia Content*. San Jose, CA, 1999.

[13] Petitcolas, F. "*Watermarking Schemes Evaluation*." *IEEE Signal Processing* 17 (2000): 58.

[14] Voloshynovskiy, S., et al. "Attack Modelling: Towards a Second Generation Benchmark." Signal Processing 81 (2001): 1177.

[15] Pereira, S., et al. "Second Generation Benchmarking and Application Oriented Evaluation." In *Proc. Information Hiding Workshop III*. Pittsburgh, PA, 2001.

[16] Solachidis, V., et al. "A Benchmarking Protocol for Watermarking Methods." In *Proc. IEEE Int. Conf. on Image Processing*. Thessaloniki, Greece, 2001.

6 Combinational Digital Watermarking in the Spatial and Frequency Domains

Digital watermarking plays a critical role in current state-of-the-art information hiding and security. It allows the embedding of an imperceptible watermark in multimedia data to identify the ownership, trace authorized users, and detect malicious attacks. Previous research posits that the embedding of a watermark into least significant bits or low-frequency components is reliable and robust. We can therefore categorize digital watermarking techniques into two embedding domains: the spatial domain and the frequency domain.

In the spatial domain we can simply insert a watermark into a host image by changing the gray levels of some pixels in the host image. The scheme is simple and easy to implement, but tends not to be robust against attacks. In the frequency domain we can insert a watermark into the coefficients of a transformed image using, for example, discrete cosine transform (DCT), discrete Fourier transform (DFT), and discrete wavelet transform (DWT). This scheme is generally considered to be robust against attacks. However, one cannot embed a large degree of watermarking in the frequency domain since the image quality after watermarking would be degraded significantly.

To provide high-capacity watermarks and minimize image distortion, this chapter presents the technique of combinational watermaking in both the spatial and frequency domains. The idea is to split the watermark image into two parts, which are respectively used for spatial and frequency insertions based on the user preference and data importance. The splitting strategy can be designed even more complicated to be unable to compose. Furthermore a random permutation of the watermark is used to enhance robustness when it is attached by some image processing operations, such as image cropping.

This chapter is organized as follows: section 6.1 presents an overview of combinational watermarking; sections 6.2 and 6.3 introduce the techniques of embedding watermarks in the spatial and frequency domains, respectively; experimental results are given in section 6.4; and section 6.5 discusses the further encryption of combinational watermarking.

6.1 AN OVERVIEW OF COMBINATIONAL WATERMARKING

In order to insert more watermarking data into a host image, the simple way is to embed it in the spatial domain of the host image. However, the disadvantage is that the inserted data could be detectable by some simple extraction skills. How can we insert a greater number of signals while still assuring that their visual effect remains visually imperceptible? The answer is a new strategy of embedding a high volume of watermarking into a host image by splitting the watermark image into two parts: one is embedded in the host image's spatial domain, and the other is embedded in the frequency domain.

If H is the original gray-scale host image of size $N \times N$ and W is the binary watermark image of size $M \times M$, let W^1 and W^2 denote the two separated watermarks from W, and H^S denote the image combined from H and W^1 in the spatial domain. H^{DCT} is the image where H^S is transformed into the frequency domain by DCT. H^F is the image where H^{DCT} and W^2 are combined in the frequency domain. Let $\oplus$ denote the operation that substitutes bits of watermark for the least significant bits (LSBs) of the host image.

The algorithm of the combinational image watermarking is presented below, and its flowchart is shown in Figure 6.1.

COMBINATIONAL WATERMARKING ALGORITHM

1. Separate the watermark into two parts:

   $W = \{w(i, j), 0 \le i, j < M\}$, where $w(i, j) \in \{0,1\}$

   $W^1 = \{w^1(i, j), 0 \le i, j < M_1\}$, where $w^1(i, j) \in \{0,1\}$

   $W^2 = \{w^2(i, j), 0 \le i, j < M_2\}$, where $w^2(i, j) \in \{0,1\}$

   $M = M_1 + M_2$

2. Insert W^1 into the spatial domain of H to obtain H^S as

   $H^S = \{h^S(i, j) = h(i, j) \oplus w^1(i, j), 0 \le i, j < N\}$, where

   $h(i, j)$ and $h^S(i, j) \in \{0,1,2,\ldots,2^L - 1\}$, and L is the number of bits used in the gray level of the pixels.

3. Transform H^S by DCT to obtain H^{DCT}.

4. Insert W^2 into the coefficients of H^{DCT} to obtain H^F as

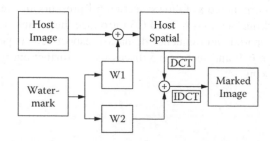

FIGURE 6.1 The flowchart in the combinationed spatial and frequency domains.

$$H^F = \{h^F(i,j) = h^{DCT}(i,j) \oplus w^2(i,j), 0 \le i,j < N\}, \text{ where}$$

$$h^F(i,j) \in \{0,1,2,\ldots,2^L - 1\}$$

5. Transform the embedded host image by Inverse DCT.

The process of splitting the watermark image into two parts, which are individually inserted into the input image in the spatial and frequency domains, depend on user requirements and applications. In principle, the most important information appears at the center of an image. Therefore, a simple way of splitting is to select the central window in the watermark image for insertion into the frequency domain. According to user preference we can crop the most private data for insertion into the frequency domain.

6.2 WATERMARKING IN THE SPATIAL DOMAIN

There are many ways of embedding a watermark into the spatial domain of a host image—for example, substituting the LSBs of some pixels [1], changing the paired pixels [2], and coding by textured blocks [3]. The most straightforward method uses the LSBs. Given the sufficiently high channel capacity in data transmission, a smaller object may be embedded multiple times. Thus, even if most of these watermarks are lost or damaged due to malicious attacks, a single surviving watermark would be considered as a success.

Despite its simplicity, the LSB substitution suffers a major drawback. Any noise addition or lossy compression is likely to defeat the watermark. An even simple attack is to simply set the LSB bits of each pixel to one. Moreover, the LSB insertion can be altered by an attacker without there being a noticeable change. An improved method would be to apply a pseudorandom number generator to determine the pixels to be used for embedding based on a designed key.

As shown in Figure 6.2, the watermarking can be implemented by modifying the bits of some pixels in the host image. Let H^* be the watermarked image. The algorithm is presented below.

Spatial Domain Algorithm

1. Obtain pixels from the host image.

$$H = \{h(i,j), 0 \le i,j < N\}, h(i,j) \in \{0,1,2,\ldots,2^L - 1\}$$

2. Obtain pixels from the watermark.

$$W = \{w(i,j), 0 \le i,j < M\}$$

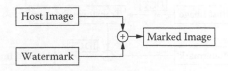

FIGURE 6.2 The flowchart in the spatial domain.

3. Substitute the pixels of the watermark into the LSB pixels of the host image.

$$H^* = \{h^*(i,j) = h(i,j) \oplus w(i,j), 0 \le i,j < N\}, h^*(i,j) \in \{0,1,2,\ldots,2^L - 1\}$$

6.3 WATERMARKING IN THE FREQUENCY DOMAIN

Watermarking can be applied in the frequency domain by first applying a transform. Similar to spatial-domain watermarking, the values of specific frequencies can be modified from the original. Since high-frequency components are often lost by compression or scaling, the watermark is embedded into lower-frequency components, or alternatively embedded adaptively into the frequency components that include critical information of the image. Since the watermarks applied to the frequency domain will be dispersed entirely over all of the spatial image after the inverse transform, this method is not as susceptible to defeat by cropping as spatial-domain approaches are.

Several approaches can be used in frequency-domain watermarking—for example, JPEG-based [4], spread spectrum [5, 6], and content-based approaches [7]. The often-used transformation functions are the DCT, DFT, and DWT, which allow an image to be separated into different frequency bands, making it much easier to embed a watermark by selecting the frequency bands of an image. One can avoid the most important visual information in the image (i.e., low frequencies) without overexposing them for removal through compression and noise attacks (i.e., high frequencies).

Generally, we can insert the watermark into the coefficients of a transformed image, as shown in Figures 6.3 and 6.4. The important consideration is what locations are the best place for embedding a watermark in the frequency domain to avoid distortion [8].

Let H^m and W^n be the subdivided images from H and W, respectively, H^{m-DCT} be the image transformed from H^m by DCT, and H^{m-F} be the image combined by H^{m-DCT} and W^n in the frequency domain. The algorithm is described as follows.

FREQUENCY DOMAIN ALGORITHM

1. Divide the host image into a set of 8×8 blocks.

    $$H = \{h(i,j), 0 \le i,j < N\}$$

    $H^m = \{h^m(i,j), 0 \le i,j < 8\}$, where $h^m(i,j) \in \{0,1,2,\ldots,2^L - 1\}$ and m is the total number of the 8×8 blocks.
2. Divide the watermark image into a set of 2×2 blocks.

    $$W = \{w(i,j), 0 \le i,j < M\}$$

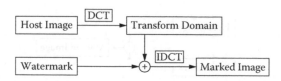

FIGURE 6.3 The flowchart in the frequency domain.

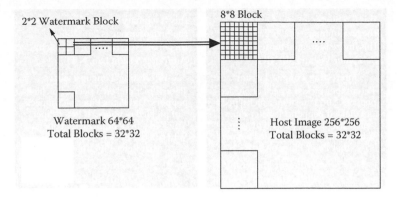

FIGURE 6.4 The embedding skill in the frequency domain.

$W^n = \{w^n(i,j), 0 \le i, j < 2\}$, where $w^n(i,j) \in \{0,1\}$ and n is the total number of the 2×2 blocks.

3. Transform H^m to H^{m_DCT} by DCT.

3. Insert W^m into the coefficients of H^{m_DCT}.

$H^{m\_F} = \{h^{m\_F}(i,j) = h^{m\_DCT}(i,j) \oplus w^m(i,j), 0 \le i, j < 8\}$, where

$h^{m\_DCT}(i,j) \in \{0,1,2,\ldots,2^L - 1\}$

5. Transform the embedded host image, H^{m_F}, by inverse DCT.

The criterion for embedding the watermark image into the frequency domain of a host image is that the total number of 8×8 blocks in the host image must be larger than the total number of 2×2 blocks in the watermark image.

6.4 EXPERIMENTAL RESULTS

Figure 6.5 illustrates that it is important to split some parts of a watermark for security purposes, such as not allowing the public to view, in the images, who the writer is. Therefore, part of the watermark is embedded into the frequency domain and the rest is embedded into the spatial domain. In this way we not only enlarge the capacity, but also secure the information concerned.

Figure 6.6 shows an original Lena image of size 256×256. Figure 6.7 is the traditional watermarking technique of embedding a 64×64 watermark into the frequency domain of a host image. Figure 6.7(a) is the original 64×64 watermark image and Figure 6.7(b) is the watermarked Lena image by embedding Figure 6.7(a) into the frequency domain of Figure 6.6. Figure 6.7(c) is the extracted watermark image from 6.7(b).

Figure 6.8 demonstrates the embedding of a large watermark, a 128×128 image, into a host image. Figure 6.8(a) is the original 128×128 watermark image, and Figures 6.8(b) and 6.8(c) are the two divided images from the original watermark, respectively. We obtain Figure 6.8(e) by embedding Figure 6.8(b) into the spatial

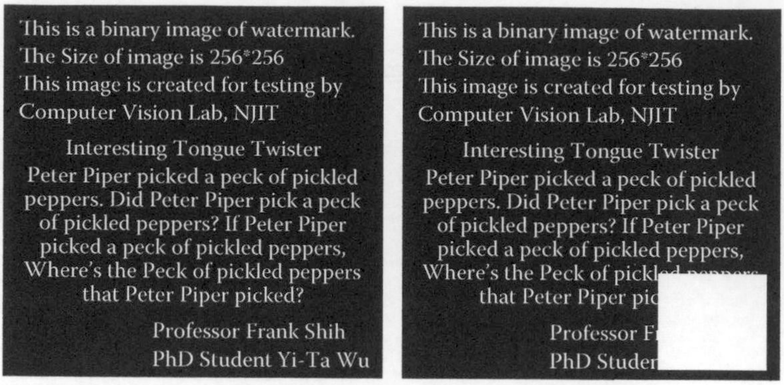

This is a binary image of watermark.
The Size of image is 256*256
This image is created for testing by
Computer Vision Lab, NJIT

Interesting Tongue Twister
Peter Piper picked a peck of pickled
peppers. Did Peter Piper pick a peck
of pickled peppers? If Peter Piper
picked a peck of pickled peppers,
Where's the Peck of pickled peppers
that Peter Piper picked?

Professor Frank Shih
PhD Student Yi-Ta Wu

This is a binary image of watermark.
The Size of image is 256*256
This image is created for testing by
Computer Vision Lab, NJIT

Interesting Tongue Twister
Peter Piper picked a peck of pickled
peppers. Did Peter Piper pick a peck
of pickled peppers? If Peter Piper
picked a peck of pickled peppers,
Where's the Peck of pickled peppers
that Peter Piper pic

Professor Fi
PhD Studer

FIGURE 6.5 A 64×64 square area is cut from a 256×256 image.

FIGURE 6.6 A Lena image.

FIGURE 6.7 The traditional technique embedding a 64×64 watermark into a Lena image.

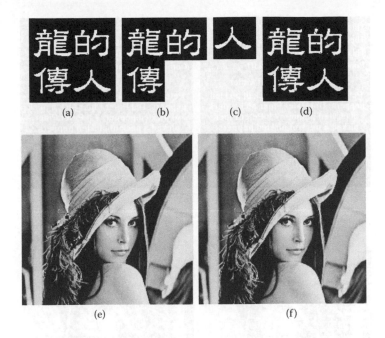

FIGURE 6.8 The results when embedding a 128×128 watermark into a Lena image.

domain of the Lena image, and obtain Figure 6.8(f) by embedding Figure 6.8(c) into the frequency domain of the watermarked Lena image in 6.8(e). Figure 6.8(d) is the extracted watermark from 6.8(f).

Figure 6.9(a) shows a larger watermark, a 256×256 image, which is split into two parts, as in Figure 6.5. Figures 6.9(c) and 6.9(d) are the watermarked images after embedding the watermark in the spatial and frequency domains of a host image, respectively. Figure 6.9(b) is the extracted watermark from Figure 6.9(d).

We can apply error measures such as *normalized correlation* (NC) and *peak signal-to-noise ratio* (PSNR) to compute the image distortion after watermarking. The correlation between two images is often used in feature detection. Normalized correlation can be used to locate a pattern on a target image that best matches the specified reference pattern from the registered image base. If $h(i, j)$ denotes the original image and $h^*(i, j)$ denotes the modified image, the normalized correlation is defined as

$$NC = \frac{\sum_{i=1}^{N} \sum_{j=1}^{N} h(i, j) h^*(i, j)}{\sum_{i=1}^{N} \sum_{j=1}^{N} [h(i, j)]^2} \tag{6.1}$$

The PSNR is often used in engineering to measure the signal ratio between the maximum power and the power of corrupting noise. Because signals possess a largely wide dynamic range, we apply the logarithmic decibel scale to limit its variation. This measures the quality of reconstruction in image compression. However, it

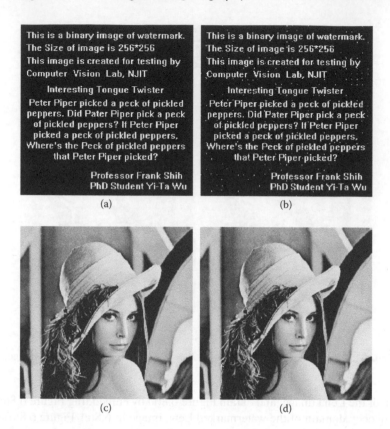

FIGURE 6.9 The results when embedding a 256×256 watermark into a Lena image.

is a rough quality measure. In comparing two video files, we can calculate the mean PSNR. The PSNR is defined thus:

$$PSNR = 10\log_{10}\left(\frac{\sum_{i=1}^{N}\sum_{j=1}^{N}[h^{*}(i,j)]^{2}}{\sum_{i=1}^{N}\sum_{j=1}^{N}[h(i,j)-h^{*}(i,j)]^{2}} \right) \tag{6.2}$$

Table 6.1 presents the results of embedding different sizes of watermarks into a host image. The PSNR in the first step indicates comparing both the original and the

TABLE 6.1

Comparisons When Embedding Watermarks of Different Sizes into a Lena Image of Size 256×256

	64×64	128×128	256×256
PSNR in First Step	none	56.58	51.14
PSNR in Second Step	64.57	55.93	50.98
NC	1	0.9813	0.9644

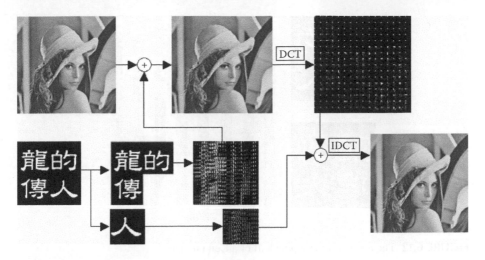

FIGURE 6.10 The procedure of randomly permuting the watermark.

embedded Lena images in the spatial domain. The PSNR in the second step indicates comparing both images in the frequency domain. The NC indicates comparing both the original and the extracted watermarks.

6.5 FURTHER ENCRYPTION OF COMBINATIONAL WATERMARKING

For the purpose of enhancing robustness, a random permutation of the watermark is used to defeat the attacks of image processing operations such as cropping. This procedure is illustrated in Figure 6.10.

A random-sequence generator is used to relocate the order of sequential numbers [9]. For example, the 12-bit random sequence generator is used to relocate the order of a watermark of size 64×64, as shown in Figure 6.11. We relocate the bits 9 and 6 at the rear of the whole sequence, and the result is shown in Figure 6.11(b).

Figure 6.12 shows the result when a half of the Lena image is cropped, where Figure 6.12(a) is the original 128×128 watermark, Figure 6.12(b) is the cropped Lena image, and Figure 6.12(c) is the extracted watermark.

Table 6.2 shows the results when parts of an embedded host image with a watermark of 128×128 are cropped. For example, if the size of 16×256 is cropped from

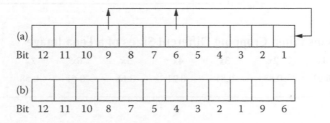

FIGURE 6.11 A 12-bit random-sequence generator.

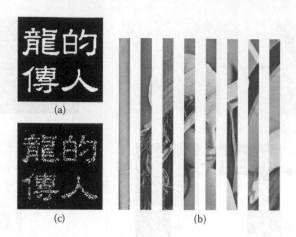

FIGURE 6.12 The result after cropping half of a Lena image.

a 256×256 embedded host image, the NC is 0.92. If a half of the embedded host image is cropped (i.e., 8 of 16×256), as shown in Figure 6.12(b), the NC is 0.52.

For the most part, the important part of an image is not enormous. We can cut the important part and embed it into the frequency domain and the rest into the spatial domain of a host image. Therefore, we cannot only enlarge the size of watermark but also retain the properties of security and imperceptibility. Combinational image watermarking possesses the following advantages [10]: more watermarking data can be inserted into the host image, so capacity is increased; the splitting of the watermark into two parts doubles the degree of protection; and the splitting strategy can be designed to involve multiple parts and their mixture to be unable to compose them.

Another scheme, that of combining spatial-domain watermarking with wavelet-domain watermarking, can be referred to in Tsai, Yu, and Chen [11]. Employing error-control codes can increase the robustness of spatial-domain watermarking [12], but the watermarking capacity is reduced since the error-control codes add some redundancy to the watermark.

TABLE 6.2
Comparisons When Cropping Different Sizes of a Lena Image

	1x	2x	3x	4x	5x	6x	7x	8x
NC	0.92	0.88	0.81	0.73	0.68	0.61	0.58	0.52

REFERENCES

[1] Wolfgang, R., and E. Delp. "A Watermarking Technique for Digital Imagery: Further Studies." In *Proc. Int. Conf. Imaging Science, Systems and Technology.* Las Vegas, NV, 1997.

[2] Pitas, I., and T. Kaskalis. "Applying Signatures on Digital Images." In *Proc. IEEE Workshop Nonlinear Signal and Image Processing.* 1995.

[3] Caronni, G. "Assuring Ownership Rights for Digital Images." In *Proc. Reliable IT Systems.* Germany: Viewveg, 1995.

[4] Zhao, K. E. "Embedding Robust Labels into Images for Copyright Protection." Technical report. Darmstadt, Germany: Fraunhofer Institute for Computer Graphics, 1994.

[5] Cox, I., et al. "Secure Spread Spectrum Watermarking for Images, Audio, and Video." In *Proc. IEEE Int. Conf. Image Processing.* Lausanne, Switzerland: IEEE, 1996.

[6] Cox, I., et al. "Secure Spread Spectrum Watermarking for Multimedia." *IEEE Trans. Image Processing* 6 (1997): 1673.

[7] Bas, P., J.-M. Chassery, and B. Macq. "Image Watermarking: An Evolution to Content Based Approaches." *Pattern Recognition* 35 (2002): 545.

[8] Hsu, C.-T. and Wu, J.-L. "Hidden Digital Watermarks in Images." *IEEE Trans. Image Processing* 8 (1999): 58.

[9] Lin, S. D., and C.-F. Chen. "A Robust DCT-Based Watermarking for Copyright Protection." *IEEE Trans. Consumer Electronics* 46 (2000): 415.

[10] Shih, F. Y., and S. Y. Wu. "Combinational Image Watermarking in the Spatial and Frequency Domains." *Pattern Recognition* 36 (2003): 969.

[11] Tsai, M. J., K. Y. Yu, and Y. Z. Chen. "Joint Wavelet and Spatial Transformation for Digital Watermarking." *IEEE Trans. Consumer Electronics* 46 (2000): 241.

[12] Lancini, R., F. Mapelli, and S. Tubaro. "A Robust Video Watermarking Technique for Compression and Transcoding Processing." In *Proc. IEEE Int. Conf. Multimedia and Expo.* Lausanne: Switzerland, 2002.

7 Genetic Algorithm- Based Digital Watermarking

In digital watermarking, the substitution approach is a popular method for embedding watermarks in an image. In order to minimize image quality degradation, least-significant-bit (LSB) substitution is commonly used. In general, LSB substitution approach works well with cryptographic technology such as Message Digest 5 (MD5) and the Rivest–Shamir–Adleman (RSA) public-key encryption algorithm in the spatial domain [1, 2]. However, LSB substitution may fail in the frequency domain approach due to rounding errors as described below.

As illustrated in Figure 7.1, a watermark is embedded into discrete cosine transform (DCT) coefficients of a cover image to generate a frequency-domain watermarked image by using the substitution embedding approach. After performing the inverse DCT (IDCT) on the frequency-domain watermarked image, the spatial-domain watermarked image—called a *stego-image*—in which most values are real numbers, can be obtained. Since the pixel intensity is stored in the integer format, the real numbers have to be converted into integers. In the literature on the subject, a rounding operation is usually recommended to convert real numbers to integers [3]. However, the question is raised whether the watermark can be correctly extracted from the stego-image. Unfortunately, the extracted watermark seems like a noisy image. Therefore, it is difficult to correctly retrieve the embedded data if the rounding technique is adopted as the substitution approach, especially LSB substitution, in the frequency domain.

This chapter briefly introduces the genetic algorithm (GA) in section 7.1 and presents the concept of GA-based watermarking in section 7.2. The GA-based rounding-error watermarking scheme is described in section 7.3, and the application of GA methodology to medical image watermarking is explored in section 7.4.

7.1 INTRODUCTION TO THE GENETIC ALGORITHM

The genetic algorithm, introduced by Hollard in his seminal work [4], is commonly used as an adaptive approach to providing randomized, parallel, and global searching based on the mechanics of natural selection and natural genetics in order to find solutions to a problem. There are four distinctions between normal optimization and search procedures: (1) GAs work with a coded parameter set, not the parameters themselves; (2) GAs search from randomly selected points, not from a single point; (3) GAs use objective function information; and (4) GAs use probabilistic transition rules, not deterministic ones [5].

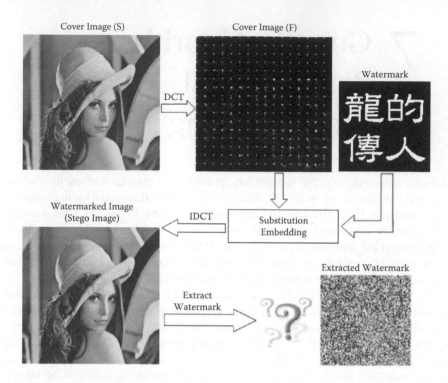

FIGURE 7.1 An example of the rounding-error problem caused by the substitution embedding approach in the frequency domain.

Although many kinds of GAs have been developed [6, 7], the fundamental of GAs is based on the simple genetic algorithm [8]. In general, GAs start with some randomly selected population, called the first generation. Each individual (called a *chromosome*) in the population corresponds to a solution in the problem domain. An objective, called *fitness function*, is used to evaluate the quality of each chromosome. The next generation will be generated from some chromosomes whose fitness values are high. Reproduction, crossover, and mutation are the three basic operators used to reproduce chromosomes in order to obtain the best solution. The process will be repeated many times until a predefined condition is satisfied or the desired number of iterations is reached. Detailed descriptions of chromosomes, operations, and fitness functions will be presented below.

7.1.1 THE CHROMOSOME

If the problem is considered to be a lock, the chromosomes can be considered keys. That is, each chromosome can be considered as a possible solution to the problem. Therefore, GAs can be considered as the procedures for selecting the correct key to open the lock, as shown in Figure 7.2.

Usually, it is possible that there is no match among the existing keys. That is, no key can be used to open the lock. In this case, the locksmith through his experience

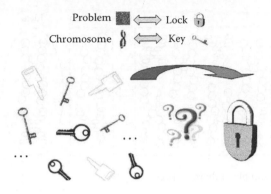

FIGURE 7.2 The procedure of selecting the correct key to open the lock.

will select a candidate key and adjust the shape of the key to generate the correct one to open the lock. Note that the criterion of selecting the candidate key is based on the adjustment complexity—that is, the smaller the candidate key adjustment, the higher the candidate key selection The same logic can be adopted in GAs. The chromosomes of the new generation are obtained based on the high-quality chromosomes of the previous generation. (The procedure of evaluating the chromosome will be presented in section 7.1.3.) An example of generating the desired chromosome through GA operations is shown in Figure 7.3. The three GA operations—reproduction, crossover, and mutation—are used to adjust the candidate chromosomes and

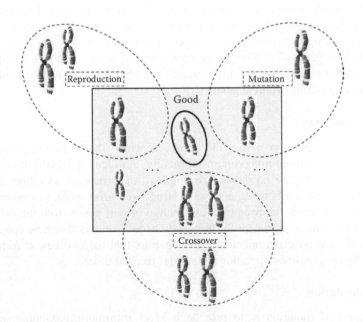

FIGURE 7.3 An example of generating the desired chromosome via the GA operation.

GA Operation—Reproduction   Ideal

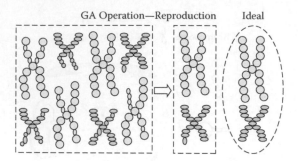

FIGURE 7.4   An example of reproduction.

generate new chromosome. Note that the procedure of creating the new generation will not be terminated until the desired chromosome is obtained.

7.1.2   BASIC OPERATIONS OF THE GENETIC ALGORITHM

The basic operations in genetic algorithm such as reproduction, crossover, and mutation are described below.

7.1.2.1   Reproduction

The purpose of reproduction is to avoid blindly searching for the desired solution by ensuring high-quality chromosomes—those that are so close to the ideal chromosomes in the current generation will be retained in the next generation. Figure 7.4 shows an example of the reproduction operation. Let the number of chromosomes in the current generation be 8 and the maintaining ratio be 0.25 (2 chromosomes will be retained among 8 chromosomes). Suppose the requirement of the ideal chromosome is that the genes of a chromosome should belong to the same size and the same shape. It is obvious that the two chromosomes are obtained after reproduction, since each of them contains different genes.

7.1.2.2   Crossover

The purpose of crossover is to avoid a case in which the searching procedure is blocked in local minimum/maximum subsets by significantly modifying the genes of a chromosome. Figure 7.5 shows an example of the crossover operation. The procedure of crossover starts by randomly selecting two chromosomes as parents from the current generation, followed by the exchange of part genes from the two parent chromosomes. The requirement for the ideal chromosome is that it be the same as shown in the reproduction operation. It is obvious that the two new chromosomes obtained by the crossover operation are similar to each other.

7.1.2.3   Mutation

The purpose of mutation is to provide a local minimum/maximum searching approach by slightly modifying the gene of a chromosome. The mutation operation

GA Operation—Reproduction

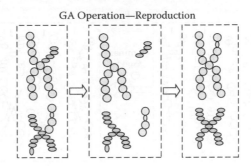

FIGURE 7.5 An example of crossover.

works especially well in cases in which the difference between an existing chromosome and a desired one is very small. For example, in Figure 7.6, the difference between the two chromosomes and the desired ones are a factor of 2. After mutation operations, four chromosomes obtained as the new generation are close to the desired ones, since the number of difference is decreased to be a factor of 1.

7.1.3 The Fitness Function

The fitness function is used to evaluate the quality of the chromosomes in the current generation. Usually the fitness function utilizes the threshold value to distinguish the good chromosomes from the bad ones, and is designed based on the problem. For example, based on the requirement as shown in the reproduction operation, the fitness function will be defined as the number of genes that do not belong to the dominant gene. The procedure to determine the dominant gene is as follows: (1) like genes among a chromosome are collected together; (2) the numbers of the genes in each category are obtained; (3) the gene with the largest number will be determined as the dominant gene. Figure 7.7 shows an example of the fitness function. There are six chromosomes in the original generation, and it is obvious that there are two kinds of

GA Operation—Mutation

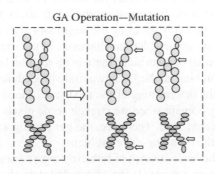

FIGURE 7.6 An example of mutation.

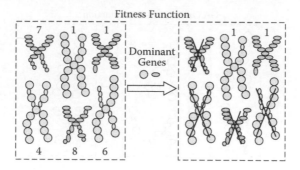

FIGURE 7.7   An example of the fitness function.

dominant gene among them. The value along each chromosome is determined by the total number of nondominant genes of a chromosome. Now, let the threshold value be 2. Therefore, there are clearly only two chromosomes that can be considered good enough to pass the evaluation of the fitness function.

7.2   THE CONCEPT OF GENETIC ALGORITHM-BASED WATERMARKING

As has been mentioned, it is possible to get a noisy watermark from a watermarked image if the LSB substitution approach is utilized as the embedding strategy and the watermark is embedded into the frequency domain. Therefore, is it possible to correct the error by changing the value of certain locations of the spatial domain? The answer is *yes*, but it is not easy, because it is very difficult to simply predict what impacts will happen in the frequency domain of a cover image if the changes are made to some values of the pixels in the spatial domain of the cover image.

In order to solve the problem, evolutionary algorithms are utilized, such as *simulated annealing*, *hill climbing*, and the genetic algorithm. The terminology comes from annealing in metallurgy, which is a technique using controlled heating and cooling of a material to increase its strength and reduce its defect. Simulated annealing has been applied in various combinatorial optimization problems and circuit design problems. In simulated annealing, the choice of an initial temperature and the corresponding temperature decrement strategy will affect performance dramatically. The procedure of randomly selecting the neighbors of a candidate in simulated annealing is to find the optimum solution. However, in the rounding-error problem, it is not easy to decide on all of them since there is no relation between a pixel and its neighbors. In the hill-climbing method, the two main procedures are the determining of the initial state and the determining of the next state. Since relations between an image and its frequency domain when we change some pixels are too complicated, it is not easy to find the next state. Therefore, the genetic algorithm is considered as the method for solving the rounding-error problem.

Before presenting the basic concept of reducing errors via genetic algorithm, we must consider two primary issues: (1) the embedded data should be accurate; and (2) changes made to modulate errors should be minimal.

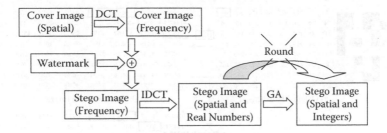

FIGURE 7.8 The purpose of the genetic algorithm.

The coefficients in the frequency domain will be changed dramatically even though only one pixel in the spatial domain is changed. Moreover, the change of pixels in the frequency domain is difficult to derive intuitively. Therefore, it is difficult to determine whether the embedded data are variant or not in the process of translating real numbers into integers. On the other hand, it is also difficult to predict and change the proper pixels in order to correct the errors and then obtain the exact embedded data. Therefore, it is not only difficult to modulate intuitively, but also probable that the worst result will be caused by random modulation.

In order to solve these problems, genetic algorithm is utilized instead of the traditional rounding approach to find a suitable solution for translating real numbers into integers, as shown in Figure 7.8. The GA-based novel method not only restores the embedded data exactly but also changes the fewest pixels in the cover image.

7.3 GENETIC ALGORITHM-BASED ROUNDING-ERROR CORRECTION WATERMARKING

The simple rounding strategy is often used for converting the real numbers into integers. However, the rounding-error problem as mentioned earlier will cause problems such that the embedded watermark is significantly destroyed. An illustration of the rounding-error problem is shown in Figure 7.9.

Figure 7.9(a) is the original host image—an 8×8 gray-scale image—in the spatial domain; Figure 7.9(b) is its transformed image by DCT. Figure 7.9(c) is a binary watermark, in which 0 and 1 denote the embedded value in its location; the minus sign (–) indicates no change in its position. Figure 7.9(d) is obtained by embedding Figure 7.9(c) into Figure 7.9(b) based on LSB modification. Note that the watermark is embedded into the integer part of the absolute real number. After transforming Figure 7.9(d) into its spatial domain via IDCT, the watermarked (stego-) image is generated as shown in Figure 7.9(e), where all pixels are real numbers. After rounding the real numbers into integers in Figure 7.9(e), the watermarked image is obtained in Figure 7.9(f).

The method for extracting the watermarks is described below. The watermarks can be extracted from the coefficients of the frequency domain of the watermarked image, Figure 7.9(f), by reperforming the embedding procedure. Therefore, Figure 7.9(f) is transformed via DCT to obtain Figure 7.9(g). Finally, the embedded

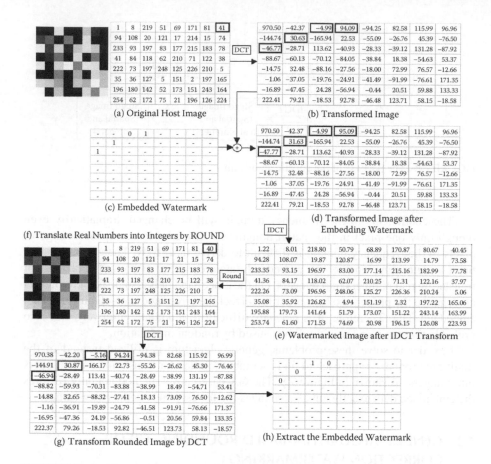

(a) Original Host Image

(b) Transformed Image

(c) Embedded Watermark

(d) Transformed Image after Embedding Watermark

(f) Translate Real Numbers into Integers by ROUND

(e) Watermarked Image after IDCT Transform

(g) Transform Rounded Image by DCT

(h) Extract the Embedded Watermark

FIGURE 7.9 An illustration of the rounding-error problem.

watermark is obtained, as shown in Figure 7.9(h) by extracting data from the specific positions in Figure 7.9(g). For example, the integer part of an absolute value of −47.94 is 47. The binary string 00101110 is obtained by translating the decimal value 46 into its binary format. Then, the value 0 is obtained from the LSB of 00101110. Similarly, a 4-bit watermark (0010) can be determined from (−47.94,30.87,−5.16,94.24). Comparing the embedded (1101) and extracted (0010) watermarks, it is found that the watermark is totally different an undesired result. In other words, the big mistake of losing the original watermark is made by using the rounding approach.

Table 7.1 shows the results when embedding different kinds of watermark into the same transformed image in Figure 7.9(d). The embedded rule is shown in Figure 7.10. That is, for a 4-bit watermark 1234, we insert the most sgnificant bit 1 into position A, 2 into position B, 3 into position C, and 4 into position D. There are 2^4 possible embedded watermarks. Only two cases can be extracted correctly.

TABLE 7.1

The Total Possible Embedded Watermarks

Embedded	Extracted	Error Bits	Embedded	Extracted	Error Bits
0000	0000	0	1000	0000	1
0001	0000	1	1001	0000	2
0010	0000	1	1010	0000	2
0011	0000	2	1011	0000	3
0100	0000	1	1100	0000	2
0101	0000	2	1101	0010	4
0110	0000	2	1110	1110	0
0111	1110	2	1111	1110	1

7.3.1 DEFINITIONS: CHROMOSOME, FITNESS FUNCTION, AND GENETIC ALGORITHM OPERATIONS

As mentioned earlier, the genetic algorithm is utilized instead of the traditional rounding approach to determine the rules for translating real numbers into integers. The definitions of *chromosome, fitness function*, and *GA operations* used in the GA-based rounding-error correction algorithm are defined herein.

7.3.1.1 Chromosome

In order to apply genetic algorithms in solving our problem, a chromosome G consisting of 64 genes is represented as $G = g_0 g_1 g_2 \ldots g_{63}$, where $g_i (0 \leq i \leq 63)$ correspond to the pixels shown in Figure 7.11. Figure 7.11(a) denotes the positions where $g_i (0 \leq i \leq 63)$ are located. For example, the chromosome G_1 shown in Figure 7.11(b), is represented as

$$G_1 = 1101010100010001001110010110111010101010100101011100101101111000 \quad (7.1)$$

-	-	C	D	-	-	-	-
-	B	-	-	-	-	-	-
A	-	-	-	-	-	-	-
-	-	-	-	-	-	-	-
-	-	-	-	-	-	-	-
-	-	-	-	-	-	-	-
-	-	-	-	-	-	-	-
-	-	-	-	-	-	-	-

FIGURE 7.10 The position where the watermark is embedded.

0	1	2	3	4	5	6	7
8	9	10	11	12	13	14	15
16	17	18	19	20	21	22	23
24	25	26	27	28	29	30	31
32	33	34	35	36	37	38	39
40	41	42	43	44	45	46	47
48	49	50	51	52	53	54	55
56	57	58	59	60	61	62	63

(a)

1	1	0	1	0	1	0	1
0	0	0	1	0	0	0	1
0	0	1	1	1	0	0	1
0	1	1	0	1	1	1	0
1	0	1	0	1	0	1	0
1	0	0	1	0	1	0	1
1	1	0	0	1	0	1	1
0	1	1	1	1	0	0	0

(b)

FIGURE 7.11 The positions corresponding to the genes.

The usage of chromosomes in GA-based watermarking is different from that of traditional methods. That is, it is not necessary to decode each chromosome's binary string into a number for evaluating its fitness value. The chromosomes are used to represent policy in translating real numbers into integers. If r is a real number, an integer r^* can be obtained by the following rules:

1. if the signal is 1, $r^* = Trunc(r) + 1$
2. if the signal is 0, $r^* = Trunc(r)$

where $Trunc(r)$ denotes the integer part of r.

An example is illustrated in Figure 7.12. Figure 7.12(a) shows an image whose pixel values are real numbers. Figure 7.12(b) is obtained by translating the real numbers into integers using the chromosomes in Figure 7.11(b).

7.3.1.2 Fitness Function

It is obvious that certain numbers of chromosomes are generated in each generation to provide possible solutions. In order to evaluate the quality of each chromosome (i.e., correctness in extracting the desired watermarks from the corresponding solution), the evaluation function is designed to calculate the differences between

1.22	8.01	218.80	50.79	68.89	170.87	80.67	40.45
94.28	108.07	19.87	120.87	16.99	213.99	14.79	73.58
233.35	93.15	196.97	83.00	177.14	215.16	182.99	77.78
41.36	84.17	118.02	62.07	210.25	71.31	122.16	37.97
222.36	73.09	196.96	248.06	125.27	226.36	210.24	5.06
35.08	35.92	126.82	4.94	151.19	2.32	197.22	165.06
195.88	179.73	141.64	51.79	173.07	151.22	243.14	16.99
253.74	61.60	171.53	74.69	20.98	196.15	126.08	223.93

(a)

2	9	218	51	68	171	80	41
94	108	19	121	16	213	14	74
233	93	197	84	178	215	182	78
41	85	119	62	211	72	123	37
233	73	197	248	126	226	211	5
36	35	126	5	151	3	197	166
196	180	141	51	174	151	244	164
253	62	172	75	21	196	126	223

(b)

FIGURE 7.12 The usage of the genes.

the two watermarks, as shown below. If ξ is a chromosome, and $Watermark^\alpha$ and $Watermark^\beta$ are the embedded and extracted watermarks, respectively, then

$$Evaluation\_1\ (\xi) = \sum_{i=0}^{all\ pixels} \left|Watermark^\alpha(i) - Watermark^\beta(i)\right| \qquad (7.2)$$

where $Watermark^\alpha(i)$ and $Watermark^\beta(i)$ denote the i-th bit of the embedded and the extracted watermarks, respectively. Note that if they are the same, $Evaluation\_1$ $1(\xi) = 0$.

7.3.1.3 Reproduction

The definition of the *reproduction* operation is

$$Reproduction\ (\xi) = \{\xi_i, \xi_i \in \Psi, Evaluation(\xi_i) \leq \Omega\} \qquad (7.3)$$

where Ω, a critical value, is used to sieve chromosomes and Ψ is the population. That is, we reproduce excellent chromosomes from the population because of their high qualities. In our experiences, we set Ω to be 10% of the total number of pixels.

7.3.1.4 Crossover

The definition of the *crossover* operation is

$$Crossover(\xi) = \{\xi_i \Theta \xi_j, \xi_i, \xi_j \in \Psi\} \qquad (7.4)$$

where Θ denotes the operation that recombines ξ by exchanging genes from their parents, ξ_i and ξ_j. This operation gestates a better offspring through the inheriting of exceptional genes from the parents. The most often used crossovers are one-point, two-point, and multipoint crossovers. When is the best time for selecting a suitable crossover? It depends on the length and structure of the chromosomes. This can be illustrated clearly by an example, as shown in Figure 7.13.

7.3.1.5 Mutation

The definition of the *mutation* operation is

$$Mutation(\xi) = \{\xi_i \circ j, \text{ ,where } 0 \leq j \leq Max\_Length(\xi_i), \xi_i \in \Psi\} \qquad (7.5)$$

		One-Point	Point		Two-Point	Point 1	Point 2
Before	Chromosome A	1011	010	Chromosome A	101111	010	0011
Crossover	Chromosome B	0010	111	Chromosome B	001011	111	1010
After	Chromosome A	1011	111	Chromosome A	101111	111	0011
Crossover	Chromosome B	0010	010	Chromosome B	001011	010	1010

FIGURE 7.13 One-point and two-point crossovers.

	One-Point		Two-Point	
Before Mutation	Chromosome A 1011010 ▲		Chromosome A 1011110100011 ▲ ▲	
After Mutation	Chromosome A 1010010 ▲		Chromosome A 1011010110011 ▲ ▲	

FIGURE 7.14 One-point and two-point mutations.

where ○ is the operation by randomly selecting chromosome ξ_i from Ψ and randomly changing bits from ξ_i. Figure 7.14 shows the one-point and two-point mutations.

7.3.2 THE GENETIC ALGORITHM-BASED ROUNDING-ERROR CORRECTION ALGORITHM

The GA-based rounding-error correction (GA-REC) algorithm is presented below. Its flowchart is shown in Figure 7.15.

THE GA-REC ALGORITHM

1. Define the fitness function, numbers of genes, sizes of population, crossover rate, critical value, and mutation rate.
2. Create the first generation by random selection.
3. Translate real numbers into integers based on each corresponding chromosome.
4. Evaluate the fitness value for each corresponding chromosome.
5. Obtain the better chromosomes based on the fitness value.
6. Recombine new chromosomes by using crossover.

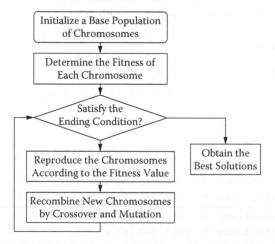

FIGURE 7.15 Basic steps in finding the best solutions through genetic algorithms.

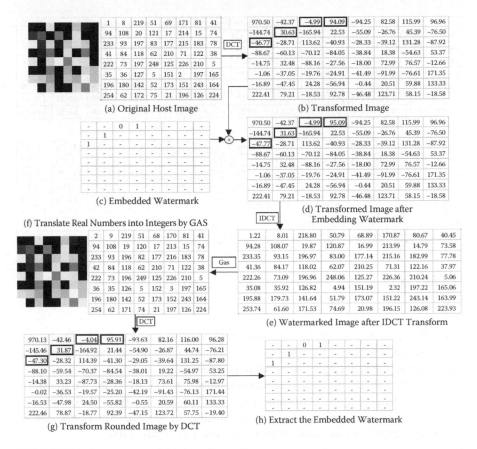

FIGURE 7.16 An example of GA-based rounding-error correction.

7. Recombine new chromosomes by using mutation.
8. Repeat steps 3–7 until a predefined condition is satisfied, or the desired number of iterations is reached.

Figure 7.16 shows the result of error correction using the genetic algorithm. Figure 7.16(a) is the original cover image, an 8×8 gray-scale image, in the spatial domain; Figure 7.16(b) is the image transformed via DCT. Figure 7.16(c) is a binary watermark, in which 0 and 1 denote the embedded data in its location; the minus sign (–) indicates no change in its position. Figure 7.16(d) is obtained by embedding Figure 7.16(c) into Figure 7.16(b) based on LSB modification. It is obvious that there are three differences when comparing Figures 7.16(b) and 7.16(d): –46.77 and –47.77, 30.63 and 31.63, and 94.09 and 95.09. After transforming Figure 7.16(d) into its spatial domain via IDCT, we obtain Figure 7.16(e), where all pixels are real numbers. After translating the real numbers into integers by GA, Figure 7.16(f) is generated.

2	9	219	50	69	170	81	41
94	108	20	120	17	214	14	74
234	93	196	82	177	216	182	78
41	85	118	62	211	71	123	37
222	74	196	248	126	226	210	5
35	36	127	5	152	2	197	165
196	180	142	51	174	151	243	163
254	61	172	74	21	196	127	223

(a)

970.25	-41.72	-4.72	95.99	-94.50	81.33	115.34	98.20
-144.70	31.52	-164.71	22.18	-54.70	-27.61	45.75	-77.27
-47.79	-28.53	114.42	-41.24	-28.69	-38.49	132.11	-86.98
-88.10	-59.77	-70.10	-84.95	-38.29	17.80	-55.40	53.56
-14.25	32.59	-87.95	-27.79	-18.50	73.65	75.12	-12.31
-0.61	-36.46	-19.83	-24.60	-41.90	-91.88	-77.15	171.89
-16.73	-47.62	24.86	-56.69	-0.02	20.89	60.33	133.01
222.11	79.33	-17.89	93.45	-45.98	123.08	58.38	-19.19

(b)

2	8	219	50	69	170	81	41
94	108	20	120	16	214	14	74
234	93	196	82	177	216	182	78
41	85	118	62	211	71	123	37
222	74	196	248	126	226	210	5
35	35	127	5	152	2	197	165
196	180	142	51	174	151	243	163
254	61	172	74	21	196	127	223

(c)

969.88	-41.98	-4.69	95.96	-94.38	81.82	115.73	98.22
-144.92	31.47	-164.56	22.08	-54.77	-27.33	45.93	-77.42
-47.96	-28.62	114.46	-41.26	-28.66	-38.27	132.28	-87.00
-88.38	-60.16	-70.32	-84.83	-37.93	18.20	-55.00	53.86
-14.13	32.56	-88.11	-27.69	-18.38	73.50	75.06	-12.13
-0.51	-36.58	-20.09	-24.45	-41.67	-91.99	-77.16	172.18
-16.80	-47.94	24.53	-56.50	0.37	21.02	60.54	133.41
222.32	79.43	-17.95	93.50	-46.00	122.80	58.18	-19.14

(d)

FIGURE 7.17 Other suitable solutions obtained by the GA-REC algorithm.

Figure 7.16(g) is the image from Figure 7.16(f) transformed via DCT. Finally, the exact embedded watermark is obtained by extracting the bits from the same position of embedding the watermark, as shown in Figure 7.16(h).

Figure 7.17 shows the other suitable solutions obtained by GAs. We transform Figure 7.17(a) into Figure 7.17(b) and Figure 7.17(c) into Figure 7.17(d). Although there are differences between Figures 7.17(b) and 7.17(d), the watermark can be exactly extracted from the same position, as shown in Figure 7.16(c).

7.3.3 An Advanced Strategy for Initializing the First Population

In Figure 7.17, although numerous suitable solutions can be obtained by GA-REC to correct the errors, many of them are useless if we consider the factor that the changes in the original image should be as few as possible. That is, the peak signal-to-noise ratio (PSNR) should be as high as possible. The definition of PSNR is

$$PSNR = 10 \times \log_{10} \left(\frac{255^2}{\sum_{i=1}^{N} \sum_{j=1}^{N} [h(i,j) - h^{GA}(i,j)]^2} \right) \quad (7.6)$$

Therefore, an improved method for minimizing the changes is developed. In order to achieve limited changes, the first population is generated based on an initial chromosome G_{Ini}, which is generated by comparing the difference between the original and rounded images. For example, there is only one difference between Figures 7.18(a) and 7.18(c), so G_{Ini} is generated as

$$G_{Ini} = 0000000100000000000000000000000000000000000000000000000000000000$$

$$(7.7)$$

1	8	219	50	69	171	81	41
94	108	20	121	17	214	15	74
233	93	197	83	177	215	183	78
41	84	118	62	210	71	122	38
222	73	197	248	125	226	210	5
35	36	127	5	151	2	197	165
196	180	142	52	173	151	243	164
254	62	172	75	21	196	126	224

(a) Original Cover Image

970.50	−42.37	−4.99	94.09	−94.25	82.58	115.99	96.96
−144.74	30.63	−165.94	22.53	−55.09	−26.76	45.39	−76.50
−46.77	−28.71	113.62	−40.93	−28.33	−39.12	131.28	−87.92
−88.67	−60.13	−70.12	−84.05	−38.84	18.38	−54.63	53.37
−14.75	32.48	−88.16	−27.56	−18.00	72.99	76.57	−12.66
−1.06	−37.05	−19.76	−24.91	−41.49	−91.99	−76.71	171.35
−16.89	−47.45	24.28	−56.94	−0.44	20.51	59.88	133.33
222.41	79.21	−18.53	92.78	−46.48	123.71	58.15	−18.58

(b) Transformed Image from (a)

1	8	219	51	69	171	81	40
94	108	20	121	17	214	15	74
233	93	197	83	177	215	183	78
41	84	118	62	210	71	122	38
222	73	197	248	125	226	210	5
35	36	127	5	151	2	197	165
196	180	142	52	173	151	243	164
254	62	172	75	21	196	126	224

(c) Translating Result by Round

970.38	−42.20	−5.16	94.24	−94.38	82.68	115.92	96.99
−144.91	30.87	−166.17	22.73	−55.26	−26.62	45.30	−76.46
−46.94	−28.49	113.41	−40.74	−28.49	−38.99	131.19	−87.88
−88.82	−59.93	−70.31	−83.88	−38.99	18.49	−54.71	53.41
−14.88	32.65	−88.32	−27.41	−18.13	73.09	76.50	−12.62
−1.16	−36.91	−19.89	−24.79	−41.58	−91.91	−76.66	171.37
−16.95	−47.36	24.19	−56.86	−0.51	20.56	59.84	133.35
222.37	79.26	−18.58	92.82	−46.51	123.73	58.13	−18.57

(d) Transformed Image from (c)

1	8	219	50	69	170	81	40
94	108	20	121	16	214	15	74
233	93	197	83	177	215	183	78
41	84	118	62	210	71	122	38
222	73	197	248	125	226	210	5
35	36	127	5	151	2	197	165
196	180	142	52	173	151	243	164
254	62	172	75	21	196	126	224

(e) Translating Result by Gas

970.13	−42.06	−4.93	93.96	−94.38	82.86	115.83	96.97
−145.23	31.04	−165.88	22.38	−55.23	−26.40	45.15	−76.46
−47.17	−28.34	113.58	−41.02	−28.39	−38.86	131.02	−87.78
−88.93	−59.83	−70.27	−84.06	−38.81	18.49	−54.92	53.63
−14.88	32.72	−88.42	−27.49	−17.88	72.98	76.27	−12.30
−1.09	−36.88	−20.06	−24.79	−41.31	−92.09	−76.88	171.73
−16.86	−47.35	24.02	−56.83	−0.28	20.39	59.67	133.66
222.44	79.26	−18.69	92.85	−46.38	123.63	58.03	−18.40

(f) Transformed Image from (e)

-	-	0	1	-	-	-	-
-	1	-	-	-	-	-	-
1	-	-	-	-	-	-	-
-	-	-	-	-	-	-	-
-	-	-	-	-	-	-	-
-	-	-	-	-	-	-	-
-	-	-	-	-	-	-	-
-	-	-	-	-	-	-	-

(g) The Extracted Data by Gas

-	-	1	0	-	-	-	-
-	0	-	-	-	-	-	-
0	-	-	-	-	-	-	-
-	-	-	-	-	-	-	-
-	-	-	-	-	-	-	-
-	-	-	-	-	-	-	-
-	-	-	-	-	-	-	-
-	-	-	-	-	-	-	-

(h) The Extracted Data by Round

FIGURE 7.18 A more suitable solution obtained by the GA-REC algorithm.

After generating G_{Ini}, the first population is generated by randomly changing limited genes in G_{Ini}. Therefore, most bits in the population will become zeros. Figure 7.18 shows the result that not only corrects the errors, but also minimizes the differences between original and translated images. Figure 7.18(a) is the original host image, an 8×8 gray-scale image, in the spatial domain; Figure 7.18(b) is the image from Figure 7.18(a) transformed via DCT. Figures 7.18(c) and 7.18(e) are the results

(a)

1	8	219	51	69	171	81	41
94	108	20	121	17	214	15	74
233	93	197	83	177	215	183	78
41	84	118	62	210	71	122	38
222	73	197	248	125	226	210	5
35	36	127	5	151	2	197	165
196	180	142	52	173	151	243	164
254	62	172	75	21	196	126	224

(b)

1	8	219	51	69	171	81	41
94	108	20	121	17	214	15	74
233	93	197	83	177	215	183	78
41	84	118	62	210	71	122	38
222	73	197	248	125	226	210	5
35	36	127	5	151	2	197	165
196	180	142	52	173	151	243	164
254	62	172	75	21	196	126	224

(c)

-	-	0	0	-	-	-	-
-	0	-	-	-	-	-	-
0	-	-	-	-	-	-	-
-	-	-	-	-	-	-	-
-	-	-	-	-	-	-	-
-	-	-	-	-	-	-	-
-	-	-	-	-	-	-	-
-	-	-	-	-	-	-	-

(d)

1	8	219	52	69	171	81	41
94	108	20	121	17	214	15	74
233	93	197	83	177	215	183	78
41	84	118	62	210	71	122	38
222	73	197	248	125	226	210	5
35	36	127	5	151	2	197	165
196	180	142	52	173	151	243	164
254	62	172	75	21	196	126	224

(e)

-	-	1	1	-	-	-	-
-	0	-	-	-	-	-	-
0	-	-	-	-	-	-	-
-	-	-	-	-	-	-	-
-	-	-	-	-	-	-	-
-	-	-	-	-	-	-	-
-	-	-	-	-	-	-	-
-	-	-	-	-	-	-	-

FIGURE 7.19 The solution when embedding 0011 via the GA-REC algorithm.

of Figure 7.18(b) transformed by via rounding and GA, respectively. Figures 7.18(d) and 7.18(f) are the images of Figures 7.18(c) and 7.18(e) transformed via DCT, respectively. Figures 7.18(g) and 7.18(h) are the extracted data from Figures 7.18(d) and 7.18(f), respectively. By comparing Figures 7.18(a) and 7.18(c), the only change, from 41 to 40, causes a problem in that we cannot extract embedded data exactly. After the best solution is found, as shown in Figure 7.18(e), an additional change, from 17 to 16, corrects the problem.

Figures 7.19 and 7.20 are other examples in correcting the embedded watermarks 0011 and 1011, respectively. Figures 7.19(a) and 7.20(a) are the original images. Figures 7.19(b) and 7.20(b) are the images transformed via rounding. Figures 7.19(c) and 7.20(c) are extracted from the coefficients of the transformed images of Figures 7.19(b) and 7.20(b), respectively. Figures 7.19(d) and 7.20(d) are the images transformed using the GA. Figures 7.19(e) and 7.20(e) are extracted from the coefficients of the transformed images of Figures 7.19(d) and 7.20(d), respectively.

Because of no difference between Figures 7.19(a) and 7.19(b), we know that the embedded watermark is changed after translating real numbers into integers. In Figures 7.19(b) and 7.19(d), the only difference, 51 versus 52, corrects the two errors of embedded data from 0000 to 0011.

We can see the same result in Figure 7.20. In Figures 7.20(a) and 7.20(b), the embedded watermark is changed by the translating procedure from real numbers to integers. Therefore, the extracted data are the same as shown in Figures 7.19(c) and 7.20(c). In Figures 7.20(b) and 7.20(d), the two differences, (118, 119 and 126, 125), correct the three errors of embedded data from 0000 to 1011.

1	8	219	51	69	171	81	41
94	108	20	121	17	214	15	74
233	93	197	83	177	215	183	78
41	84	**118**	62	210	71	122	38
222	73	197	248	125	226	210	5
35	36	127	5	151	2	197	165
196	180	142	52	173	151	243	164
254	62	172	75	21	196	**126**	224

(a)

1	8	219	51	69	171	81	41
94	108	20	121	17	214	15	74
233	93	197	83	177	215	183	78
41	84	**118**	62	210	71	122	38
222	73	197	248	125	226	210	5
35	36	127	5	151	2	197	165
196	180	142	52	173	151	243	164
254	62	172	75	21	196	**126**	224

(b)

-	-	0	0	-	-	-	-
-	0	-	-	-	-	-	-
0	-	-	-	-	-	-	-
-	-	-	-	-	-	-	-
-	-	-	-	-	-	-	-
-	-	-	-	-	-	-	-
-	-	-	-	-	-	-	-
-	-	-	-	-	-	-	-

(c)

1	8	219	51	69	171	81	41
94	108	20	121	17	214	15	74
233	93	197	83	177	215	183	78
41	84	**119**	62	210	71	122	38
222	73	197	248	125	226	210	5
35	36	127	5	151	2	197	165
196	180	142	52	173	151	243	164
254	62	172	75	21	196	**125**	224

(d)

-	-	1	1	-	-	-	-
-	0	-	-	-	-	-	-
1	-	-	-	-	-	-	-
-	-	-	-	-	-	-	-
-	-	-	-	-	-	-	-
-	-	-	-	-	-	-	-
-	-	-	-	-	-	-	-
-	-	-	-	-	-	-	-

(e)

FIGURE 7.20 The solution when embedding 1011 via the GA-REC algorithm.

7.4   AN APPLICATION FOR MEDICAL IMAGE WATERMARKING

In recent decades, with the rapid development of biomedical engineering, digital medical images have become increasingly important in hospitals and clinical environments. Concomitantly, medical images traversing between hospitals produces complicated network protocol, image compression, and security problems. Many techniques have been developed to resolve these problems. For example, the hospital information system (HIS) and the picture archiving and communication system (PACS) are currently the two primary data-communication systems used in hospitals. Although HIS may be slightly different between hospitals, data can be exchanged based on the Health Level Seven (HL7) standard. Similarly, PACS transmits medical images using the Digital Imaging and Communications in Medicine standard. Furthermore, the Institute of Electrical and Electronics Engineers standard 1073 was published in order to set a standard for measured data and signals from different medical instruments.

Some techniques, such as *HIS* and *PACS*, were developed to provide an efficient mechanism in dealing with the recent outburst of digital medical images. In order to enhance performance, we need to compress image data for efficient storage and transmission purposes. Lossless compression is adopted for avoiding any distortion when the images are needed for diagnosis. Otherwise, lossy compression can achieve a higher compression rate. Since the quality of medical images is crucial in diagnosis, lossless compression is often used. However, if we can utilize the domain knowledge of specific types of medical images, a higher compression ratio could be achieved without losing any important diagnostic information. In order to obtain the higher compression rate, a hybrid-compression method has been developed by compressing the region of interest (ROI) with lossless compression and the rest with

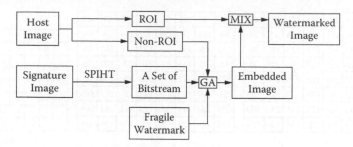

FIGURE 7.21 The encoding procedure for embedding a signature image.

lossy compression [9, 10]. In order to protect the copyright of medical images, the watermark is embedded surrounding the ROI. Meanwhile, the embedded watermark is preprocessed via set partitioning in hierarchical tree (SPIHT) for robustness [11].

7.4.1 AN OVERVIEW OF THE PROPOSED TECHNIQUE

In order to achieve a higher compression rate without distorting the important data in a medical image, we select the ROI of a medical image and compress it via lossless compression, and the rest of the image via lossy compression. For the purpose of protecting medical images and maintaining their integrity, we embed an information watermark (a signature image or textual data) and a fragile watermark into the frequency domain surrounding the ROI. Meanwhile, the embedded information watermark is preprocessed into a bitstream.

7.4.1.1 The Signature Image

7.4.1.1.1 The Encoding Procedure

Figure 7.21 shows the encoding procedure when the watermark is a signature image. The host image (medical image) is separated into two parts, the ROI and the non-ROI. We embed the signature image and fragile watermark into the non-ROI part via genetic algorithm. Note that we preprocess the signature image by SPIHT to satisfy human perception. Finally, we obtain the watermarked image by combining the embedded non-ROI and ROI.

7.4.1.1.2 The Decoding Procedure

Figure 7.22 shows the decoding procedure. The non-ROI part is selected from the watermarked image, and is transformed via DCT. Finally, we obtain the fragile watermark and a set of bitstreams by extracting data from the specific positions of the coefficients in the frequency domain. We can decide the integrity of the medical image by checking the fragile watermark, and obtain the signature image by reconstructing the bitstream using SPIHT.

7.4.1.1.3 Set Partitioning in Hierarchical Tree Compression

SPIHT is a zerotree structure based on DWT. The first zerotree structure, the embedded zerotree wavelet, was published by Shapiro in 1993 [12]. SPIHT uses a bit allocation strategy in essential and produces a progressively embedded scalable bitstream.

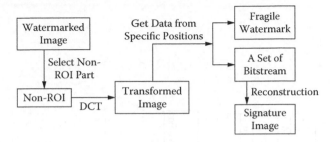

FIGURE 7.22 The decoding procedure for embedding a signature image.

Figure 7.23 shows an example of SPIHT, where Figure 7.23(a) is the original 8×8 image. We obtain the transformed image as Figure 7.23(b) via DWT. Figure 7.23(c) lists the bitstreams generated by applying SPIHT three times. We reconstruct the image by using these bitstreams, and the result is shown in Figure 7.23(d).

7.4.1.2 Textual Data

7.4.1.2.1 Encoding and Decoding Procedures

Figure 7.24 shows the encoding procedure when the watermark is textual data. The procedure is similar to signature image encoding except the encryption of textual data is used. Its decoding procedure is shown in Figure 7.25.

7.4.1.2.2 The Encryption Scheme in Textual Data

There are many techniques for encrypting textual data. Generally, the main idea is translating the textual data from plain text into secret codes of cipher text. There are two types of encryption: *asymmetric* and *symmetric*.

An easy way to achieve encryption is through bit shifting. That is, we consider a character as a byte value and shift its bit position. For example, the byte value of the letter F is 01000110. We can change it to be d as 01100100 by shifting the leftmost 4 bits to the right side.

An encryption method by taking the logarithm of the American Standard Code for Information Interchange (ASCII) has been developed by Rajendra Acharya et al. [13]. Their encryption algorithm can be mathematically stated as

$$T_e = (\log(T_o^* 2)^* 100) - 300 \tag{7.8}$$

where T_e denotes the encrypted text and T_o denotes the ASCII code of the original text. The decrypted text can be obtained by

$$T_o = \exp\left\{(T_e + 300)/100 - \log 2\right\} \tag{7.9}$$

Note that the encrypted information (T_e) is stored as an integer.

(a) Original 8*8 Image

192	192	192	192	192	192	16	16
192	16	16	192	192	192	192	192
192	16	16	192	192	192	192	192
192	16	16	16	16	192	192	192
192	16	16	192	192	192	192	192
192	16	16	192	192	192	192	192
192	16	16	192	192	192	192	192
192	192	192	192	192	192	192	192

(b) Transformed Image by DWT

139.75	−24.75	11	−11	44	0	0	−44
−2.75	275	11	−22	88	−88	0	0
11	−33	11	−11	88	−44	0	0
−33	−22	11	−22	44	−44	0	0
44	88	88	44	−44	0	0	44
0	0	0	0	0	0	0	0
0	−44	−88	0	0	44	0	0
−44	0	0	0	44	−44	0	0

(c) The Bitstreams by SPIHT

```
110000001
0000100001000001110011100110011100001000010000000
00001000010100100001111100010001010110011100111110011111111000000000000000000
```

(d) The Reconstructed Image

138.25	0	0	0	40	0	0	−40
0	0	0	0	92.5	−92.5	0	0
0	−40	0	0	92.5	−40	0	0
−40	0	0	92.5	40	−40	0	0
40	92.5	92.5	40	−40	0	0	40
0	0	0	0	0	0	0	0
0	−40	−92.5	0	0	40	0	0
−40	−40	−40	40	−40	0	0	0

FIGURE 7.23 An example of SPIHT.

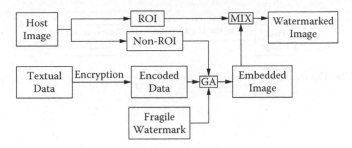

FIGURE 7.24 The encoding procedure for embedding textual data.

7.4.1.2.3 The Watermarking Algorithm for Medical Images

Let H be the original host image with size $K \times K$, which is separated into H^{ROI} (ROI) and H^{NROI} (non-ROI) with sizes of $N \times M$ and $K \times K - N \times M$, respectively. Let S and T denote a signature image with size of $W \times W$ and textual data, respectively. S^B is a bitstream obtained by compressing S in *SPIHT* compression or encoding T_o by the encryption technique. W^F is the fragile watermark. H^{WROI} is obtained by adjusting the pixel values of H^{NROI} in which we can extract S^B and W^F from some specific positions of coefficients of the frequency domain in H^{WROI}. H^{Final} is the final image by combining H^{ROI} and H^{WROI}.

THE WATERMARKING ALGORITHM FOR MEDICAL IMAGES

1. Separate the host image, H, into H^{ROI} and H^{NROI}.

   $H = \{\ h(i,j)\ ,\ 0 \le i,j < K\ \}$, where $h(i,j) \in \{0,1,2,...,2^L - 1\}$ and L is the number of bits used in the gray level of pixels.

   $H^{ROI} = \{\ h^{ROI}(i,j)\ ,\ 0 \le i < N\ ,\ 0 \le j < M\ \}$, where $h^{ROI}(i,j) \in \{0,1,2,...,2^L - 1\}$

   $H^{NROI} = \{\ h^{NROI}(i,j)\ \}$, where $h^{NROI}(i,j) \in \{0,1,2,...,2^L - 1\}$

2. (a) For the signature image: Transform S to S^B by *SPIHT*.

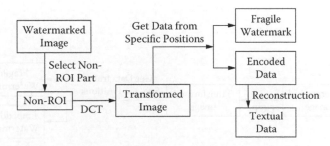

FIGURE 7.25 The decoding procedure for embedding textual data.

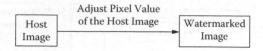

FIGURE 7.26 The encoding procedure of the improved scheme.

$S = \{\, s(i,j)\,,\ 0 \le i, j < W\,\}$, where $s(i,j) \in \{0,1,2,...,2^L - 1\}$.
S^B is the bitstream consisting of bits 0 and 1.

   (b) For the textual data: Encode T to S^B by the encryption technique.

3. Adjust H^{NROI} by genetic algorithms to obtain H^{WROI} in which we can extract S^B and W^F from the coefficients of its frequency domain.

4. Combine H^{ROI} and H^{WROI} to obtain the final image H^{Final}.

Note that we can embed not only the information image but also the fragile watermark into the host image. If there is more than one regular ROI, we can record the following information for each ROI: the top-left corner coordinates, the width, and the height. The watermark image can be placed in a group of 8 × 8 blocks that are extracted from the outer layer of each ROI. For an irregular polygon or circular ROI, we need to record the starting and ending positions of each row in ROI in order to extract the data from the outer layer of its boundary.

7.4.2 THE IMPROVED SCHEME BASED ON GENETIC ALGORITHMS

This section presents an improved scheme for embedding a watermark into the frequency domain of a host image. The new scheme not only reduces the cost of obtaining a solution but also offers more applications in watermarking. The principal idea is to adjust the pixels in the host image based on genetic algorithms, and make sure the extracted data from the specific positions in the frequency domain of the host image are the same as in the watermark. Figures 7.26 and 7.27 show the encoding and decoding procedures of the improved scheme, respectively.

The improved algorithm for embedding watermark based on genetic algorithms is presented below. Its flowchart is shown in Figure 7.28.

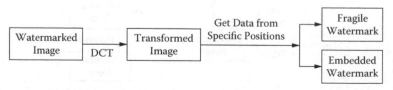

FIGURE 7.27 The decoding procedure of the improved scheme.

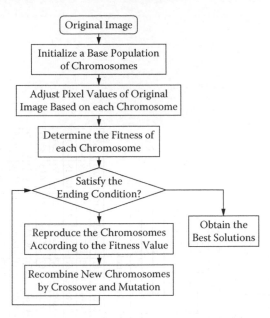

FIGURE 7.28 Basic steps in finding the best solutions by via improved GAs.

THE IMPROVED ALGORITHM

1. Define the fitness function, number of genes, size of population, crossover rate, and mutation rate.
2. Generate the first generation via random selection.
3. Adjust the pixel values based on each chromosome of the generation.
4. Evaluate the fitness value for each corresponding chromosome.
5. Obtain the better chromosomes.
6. Recombine new chromosomes by using crossover.
7. Recombine new chromosomes by using mutation.
8. Repeat steps 3 to 7 until a predefined condition is satisfied, or a constant number of iterations is reached.

Figure 7.29 shows an example of the improved genetic algorithms. Figures 7.29(a) and 7.29(b) are the original image and the signature data, respectively. Using the embedded order shown in Figure 7.29(c), we obtain Figure 7.29(d) by separating the signature data into these 12 parts. Figure 7.29(e) is the fragile watermark, which is defined by the user. Our purpose is to adjust the pixel values of Figure 7.29(a) in order to obtain its frequency domain, from which we can extract signature data and fragile watermarks from specific positions. Here the signature data and the fragile watermark are extracted from the positions of bits 3, 4, 5, and bit 1, respectively. Figure 7.29(f) shows the result of the adjusted image. We can extract the signature data and fragile watermark from the coefficients of frequency domain of the watermarked image, as shown in Figure 7.29(g). Table 7.2 shows the watermark extracted from the specific positions of the coefficients of the frequency domain of the original image.

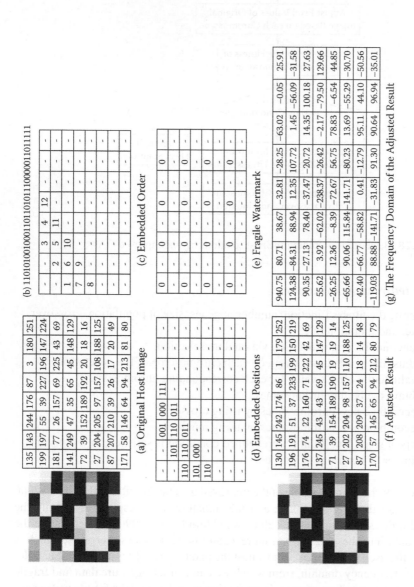

FIGURE 7.29 An example of improved genetic algorithms.

TABLE 7.2
The Watermark Extracted from the Specific Positions of the Binary Form

Decimal	Binary	Decimal	Binary	Decimal	Binary
90	010*11*010	88	010*11*000	3	00000*0*11
84	010*1*0*1*00	27	000*11*011	78	010*01*110
38	001*00*110	55	001*10*111	12	0000*11*00
32	001*0*0000	26	000*11*010	28	000*11*100

7.4.3 EXPERIMENTAL RESULTS

Figure 7.30 shows the example of embedding a signature image into a *MRI* brain image. Figures 7.30(a) and 7.30(b) show the original medical image with a size of 230×230 and the signature image with a size of 64×64, respectively. Figure 7.30(c) is the image transformed via DWT. The ROI is marked as a rectangle with size of 91×112, as shown in Figure 7.30(d). We encode Figure 7.30(c) by SPIHT into a set of bit-streams that is embedded around the ROI. In Figure 7.30(e), the area between two rectangles, 91×112 and 117×138, is the clipped watermarked area. The signature image is extracted and the reconstructed result is shown in Figure 7.30(f). Table 7.3 shows the

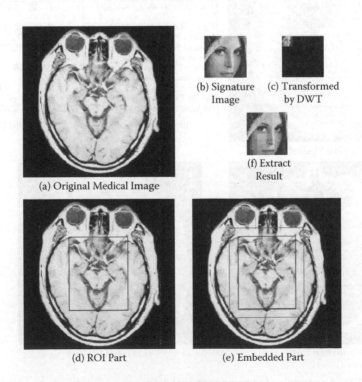

(b) Signature Image   (c) Transformed by DWT

(f) Extract Result

(a) Original Medical Image

(d) ROI Part   (e) Embedded Part

FIGURE 7.30 An example of embedding a signature image.

TABLE 7.3
The PSNR of Signature Images and Medical Images

	Original and Reconstructed Signature Images	Nonwatermarked and Watermarked Medical Images
PSNR	24.08	38.28

PSNR of the original and reconstructed signature images, and of the nonwatermarked and watermarked parts. The error measures, PSNR, is defined as follows:

$$PSNR = 10^{*}\log_{10}\left(\frac{\sum_{i=1}^{N}\sum_{j=1}^{N}[h^{GA}(i,j)]^2}{\sum_{i=1}^{N}\sum_{j=1}^{N}[h(i,j)-h^{GA}(i,j)]^2}\right) \qquad (7.10)$$

Figure 7.31 shows an example of embedding textual data into a computer tomography brain image. Figures 7.31(a) and 7.31(b) show the original medical image

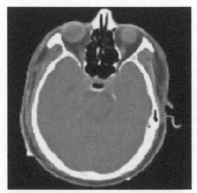

(a) Original Medical Image

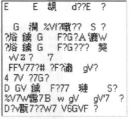

THE PATIENT INFORMATION

Patient Ref. No:AX8865098
Name of the doctor:Dr.Wu
Name of the doctor:Mr.Cheng
Age:48 years
Address:22 Midland Ave.
Case History:
Date of admission:18.05.2001
Results:T wave inversion
Diagnosis:Suspected MI

(b) Textual Data

E   E 胡   d??E ?

 G   澗 %Vf?噪?? S ?
?焓 鐹 G   F?G?A 邋W
?焓 鐹 G   F?G??? 癸
  Wз ?   7
FF√7??# ?F?浦   g√?
4 7V ??G?
D GV 鐹 F?77 楝   S?
%V7w器7B w g√   g√'7 ?
D?v覇7??W7 V6GVF ?

(c) Textual Data

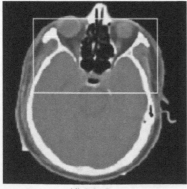

(d) ROI Part

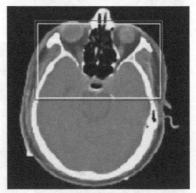

(e) Embedded Part

FIGURE 7.31 An example of embedding textual data.

with a size of 260 × 260, and the textual data, respectively. Figure 7.31(c) is the encrypted data. The ROI is marked as a rectangle with a size of 179 × 109, as shown in Figure 7.31(d). We obtain a set of bitstreams by shifting the rightmost 4 bits to the left side. In Figure 7.31(e), the area between two rectangles is the clipped watermarked area. Note that the original and extracted textual data are exactly the same.

Indeed, the genetic algorithm consumes more time. Our techniques are focused on the following two goals: (1) the embedded watermarks should be as robust as possible and can be used to detect any unauthorized modification; (2) the compression rate of an image should be as high as possible. Furthermore, the embedded watermarks will be disturbed when we convert real numbers into integers. Genetic algorithms are adopted to be the best way to achieve these goals.

In general, computational time is closely related to the amount of required embedded data. That is, the more embedded data we have, the more computational time it takes. For example, it takes about four minutes on a Pentium III PC with 600 MHz to obtaining the results shown in Figure 7.29 since there are 36 digits of bitstream and 16 digits of fragile watermark.

REFERENCES

[1] Rivest, R. L. "The MD5 Message Digest Algorithm." RFC 1321, 1992. Available online at http://www.faqs.org/rfcs/rfc1321.html.

[2] Rivest, R., A. Shamir, and L. Adleman. "A Method for Obtaining Digital Signatures and Public-Key Cryptosystems." *Communications of the ACM* 21 (1978): 120.

[3] Lin, S. D., and C.-F. Chen. "A Robust DCT-Based Watermarking for Copyright Protection." *IEEE Trans. Consumer Electron*ics 46 (2000): 415.

[4] Holland, J. H. *Adaptation in Natural and Artificial Systems.* Ann Arbor: University of Michigan Press, 1975.

[5] Herrera, F., M. Lozano, and J. L. Verdegay. "Applying Genetic Algorithms in Fuzzy Optimization Problems. *Fuzzy Systems and Artificial Intelligence* 3 (1994): 39.

[6] Ho, S.-Y., H.-M. Chen, and L.-S. Shu. "Solving Large Knowledge Base Partitioning Problems Using the Intelligent Genetic Algorithm." In *Proc. Int. Conf. Genetic and Evolutionary Computation.* Orlando: FL, 1999.

[7] Tang, K.-S., et al. "Minimal Fuzzy Memberships and Rules Using Hierarchical Genetic Algorithms." *IEEE Trans. Industrial Electronics* 45 (1998): 162.

[8] Holland, J. H. *Adaptation in Natural and Artificial Systems: An Introductory Analysis with Applications to Biology, Control, and Artificial Intelligence.* Boston: MIT Press, 1992.

[9] Wakatani, A. "Digital Watermarking for ROI Medical Images by Using Compressed Signature Image." In *Proc. Int. Conf. on System Sciences.* Kona, HI, 2002.

[10] Strom, J., and P. C. Cosman. "Medical Image Compression with Lossless Regions of Interest." *Signal Processing* 59 (1997): 155.

[11] Said, A., and W. A. Pearlmana. "A New, Fast, and Efficient Image Codec Based on Set Partitioning in Hierarchical Trees." *IEEE Trans. Circuits and System for Video Technology* 6 (1997): 243.

[12] Shapiro, J. "Embedded Image Coding Using Zerotrees of Wavelet Coefficients." *IEEE Trans. Signal Processing* 41 (1993): 3445.

[13] Acharya, U. R., et al. "Compact Storage of Medical Image with Patient Information." *IEEE Trans. Info. Tech. in Biomedicine* 5 (2001): 320.

8 Adjusted-Purpose Digital Watermarking

Several purpose-oriented watermarking techniques have been developed. Cox et al. have proposed spread-spectrum watermarking to embed a watermark into regions having large coefficients in the transformed image [1]. Wong has presented the block-based fragile watermarking technique by adopting the Rivest–Shamir–Adleman public key encryption algorithm and Message Digest 5 for the hashing function [2–4]. Celik et al. have proposed a hierarchical watermarking approach to improve Wong's algorithm [5]. Chen and Lin have presented mean quantization watermarking by encoding each bit of the watermark into a set of wavelet coefficients [6]. Note that all of these existing techniques can achieve only one purpose with watermarking (embedding either robust or fragile watermarks) and can be adapted to only one domain (either the spatial domain or the frequency domain).

Selection of a suitable watermarking technique is not easy, since there are many different kinds of watermarking techniques for variant types of watermarks and for variant purposes. In this chapter, a novel adjusted-purpose (AP) watermarking technique is developed that integrates different types of watermarking techniques. Section 8.1 presents an overview of the watermarking technique. The morphological approach for extracting pixel-based features of an image is presented in section 8.2. In section 8.3, the strategies for adjusting the varying-sized transform window (VSTW) and the quantity factor (QF) are given. Experimental results are provided in section 8.4. Section 8.5 presents a collecting approach for generating the VSTW.

8.1 AN OVERVIEW OF ADJUSTED-PURPOSE DIGITAL WATERMARKING

This section presents an overview of the adjusted-purpose digital watermarking technique. Before presenting the basic concept, two primary issues are of concern:

1. How can we adjust the approach of performing watermarking from the spatial domain to the frequency domain?
2. How can we adjust the purpose of watermarking from fragile to robust?

In order to answer these questions we develop two parameters, the VSTW and the QF. By adjusting the size of the VSTW, approaches toward performing watermarking will become a spatial-domain approach when the VSTW = 1×1 and a frequency-domain approach for other sizes. On the other hand, by adjusting the value of the QF, the embedded watermarks will become robust if the QF is large, and become fragile if the QF is 1. Following are the encoding and decoding procedures of the AP watermarking algorithm.

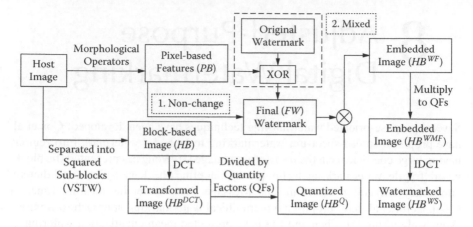

FIGURE 8.1 The encoding procedure.

Let H be a gray-level host image of size $N \times N$, and W be a binary watermark image of size $M \times M$. Let HB be the block-based image obtained by dividing H into nonoverlapping blocks of size $n \times n$, and HB^{DCT} be the image transformed via discrete cosine transform (DCT). Let Q be a quantity factor of the same size as HB, and HB^Q be the image obtained by dividing HB^{DCT} by Q. Let PB be the pixel-based features extracted from H, and FW be the final watermark by applying an *exclusive or* (XOR) operator of PB and W. Let $\otimes$ denote the operator that substitutes bits of a watermark for bits of a host image using least-significant-bit (LSB) modification. Note that the LSB modification tends to exploit the bit plane such that the modification does not cause significantly different visual perception. Let HB^{WF} be the watermarked image, where HB^Q and FW are combined using LSB modification. Let HB^{WMF} be the watermarked image obtained by multiplying HB^{WF} by Q. Let HB^{WS} be the watermarked image obtained by converting HB^{WMF} back to the spatial domain via inverse discrete cosine transform (IDCT).

The algorithm for the encoding procedure of our AP watermarking technique is presented below, and its flowchart is shown in Figure 8.1.

THE ENCODING PROCEDURE OF THE AP WATERMARKING ALGORITHM

1. Obtain PB from H based on morphological operators.
2. (a) Case 1: Without an additional watermark. Set $FW = PB$.
   (b) Case 2: With a watermark W. Obtain FW by applying the XOR operator of PB and W.
3. Determine n and obtain HB by splitting H into nonoverlapping subwatermarks of size $n \times n$.
4. Transform HB by DCT to obtain HB^{DCT}.
5. Obtain HB^Q by dividing HB^{DCT} by Q.
6. Insert FW into coefficients of HB^Q by LSB modification to obtain HB^{WF}.

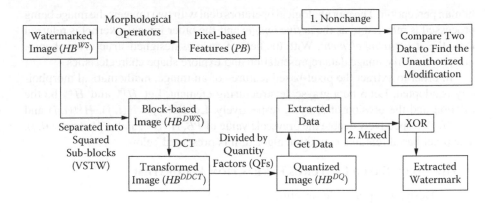

FIGURE 8.2 The decoding procedure.

7. Obtain HB^{WMF} by multiplying HB^{WF} by Q.
8. Obtain HB^{WS} by converting HB^{WMF} back to the spatial domain by IDCT.

Let HB^{DWS} be the block-based image by dividing HB^{WS} into nonoverlapping blocks of size $n \times n$. Let HB^{DDCT} be the image of HB^{DWS} transformed via DCT, and HB^{DQ} be the image obtained by dividing HB^{DDCT} by Q. Let HB^{LSB} be the binary image obtained by extracting the LSB value of each pixel in HB^{DQ}. Let FW^D be the extracted watermark by applying an XOR operator of PB and HB^{DQ}.

The algorithm for the decoding procedure of our AP watermarking technique is presented below, and its flowchart is shown in Figure 8.2.

THE DECODING PROCEDURE OF THE AP WATERMARKING ALGORITHM

1. Obtain PB from HB^{WS} based on morphological operators.
2. Determine n and obtain HB^{DWS} by splitting HB^{WS} into nonoverlapping sub-watermarks of size $n \times n$.
3. Transform HB^{DWS} via DCT to obtain HB^{DDCT}.
4. Obtain HB^{DQ} by dividing HB^{DDCT} by Q.
5. Obtain HB^{LSB} by extracting the LSB value of each pixel in HB^{DQ}.
6. (a) Case 1: Without an additional watermark. Set $FW^D = HB^{LSB}$.
   (b) Case 2: With a watermark W. Obtain FW^D by applying the XOR operator of PB and HB^{DQ}.

8.2 THE MORPHOLOGICAL APPROACH FOR EXTRACTING PIXEL-BASED FEATURES

Mathematical morphology [7–9], which is based on set-theoretic concept, can extract object features by choosing a suitable structuring shape as a probe. The analysis is geometric in character and it approaches image processing from the vantage point of human perception. The morphological operators deal with two images: the image being processed is referred to as the *active image*, and the other image, a kernel is referred to as the *structuring element*. With the help of various designed structuring shapes, one can simplify image data representation and explore shape characteristics.

In order to extract the pixel-based features of an image, mathematical morphology is adopted. Let S be a grayscale structuring element. Let H^{Di} and H^{Er} be the dilation and the erosion of H by S, respectively. Let $PB(i, j)$, $H(i, j)$, $H^{Di}(i, j)$ and $H^{Er}(i, j)$ respectively denote the grayscale value of PB, H, H^{Di}, and H^{Er} at pixel (i, j). The pixel-based features extraction algorithm is presented below.

THE PIXEL-BASED FEATURES EXTRACTION ALGORITHM

1. Design the grayscale structuring element S.
2. Obtain H^{Di} by a dilation of H by S.
3. Obtain H^{Er} by an erosion of H by S.
4. Obtain the pixel-based features by the following rule:

If $(0 \leq \frac{H(i,j)-H^{Er}(i,j)}{H^{Di}(i,j)-H^{Er}(i,j)} \leq T_1)$, or $(T_2 \leq \frac{H(i,j)-H^{Er}(i,j)}{H^{Di}(i,j)-H^{Er}(i,j)} \leq 1)$ then $PB(i, j) = 1$; otherwise, $PB(i, j) = 0$.

Note that T_1 and T_2 are threshold values, and $T_1 < T_2$.

We observe that almost all of the pixel-based features of an image will remain the same after compression. Figure 8.3 illustrates the property of pixel-based features.

82	76	53	34	33	45	59	86	72	74	124	111
73	64	46	32	54	64	80	93	102	97	87	80
59	40	34	36	38	50	68	84	96	96	91	86
53	33	28	31	24	36	53	71	86	90	88	86
62	55	42	31	30	38	51	64	69	71	70	69
58	60	55	49	44	48	53	58	57	56	53	51
44	45	59	73	53	54	55	56	51	49	47	46
38	38	53	68	50	51	52	54	52	54	56	57
39	43	43	40	43	44	47	52	58	65	71	75
55	45	63	58	46	45	44	45	49	55	62	67
55	32	68	50	36	36	37	42	51	64	75	82
51	22	67	43	30	30	32	40	54	71	88	98

(a)

83	77	50	32	38	45	61	90	70	68	127	117
69	66	48	39	49	57	71	95	102	95	88	82
53	49	35	31	42	53	67	88	103	92	951	77
54	42	27	20	26	37	55	72	82	85	86	83
64	51	39	32	30	37	52	63	66	74	70	72
61	53	55	56	47	47	55	57	54	60	53	52
47	45	60	68	55	50	56	53	48	62	30	44
39	39	59	69	54	49	56	53	54	55	59	59
44	38	41	46	41	42	51	51	54	71	69	76
55	45	64	59	46	40	46	44	46	57	56	70
53	35	69	55	39	32	40	40	56	68	74	89
50	24	61	41	28	26	37	38	53	72	88	103

(b)

82	67	48	35	36	47	61	70	61	94	109	114
70	59	45	38	44	61	80	92	95	101	86	79
51	43	32	27	34	52	73	87	98	100	85	83
49	42	32	24	24	35	51	63	82	89	86	96
62	59	51	41	35	38	48	58	75	73	63	72
56	59	59	53	46	45	53	61	57	55	47	57
39	48	56	56	49	46	51	57	44	47	48	62
35	47	60	62	55	50	52	57	56	53	46	54
41	44	46	47	45	45	48	50	57	62	61	81
47	51	55	52	45	42	44	47	51	59	61	82
46	54	58	52	40	32	34	40	52	64	68	89
37	47	54	48	34	25	27	33	57	74	81	99

(c)

88	71	46	28	27	42	64	81	68	92	117	120
82	68	49	36	39	56	80	96	87	97	107	109
63	53	39	32	38	57	79	94	103	96	90	94
45	37	27	22	27	40	56	66	96	83	74	81
48	43	35	29	29	34	41	46	72	65	64	74
60	57	53	49	47	47	48	49	52	55	63	70
50	52	53	55	56	56	56	56	49	59	69	68
28	32	39	47	52	55	56	56	54	67	75	66
52	54	56	57	55	50	46	42	44	60	73	68
49	50	52	53	51	47	42	39	45	61	75	71
44	45	48	48	46	42	37	34	46	64	79	76
40	42	44	45	43	39	34	31	49	67	84	83

(d)

FIGURE 8.3 An example of illustrating the property of pixel-based features.

TABLE 8.1
The Position of a Pixel and Its Neighborhood in Different Compression Rates

Figure	Pixel Value	Sorted Sequence	Order
8.3a	34	28,31,32,33,*34*,36,40,46,64	5
8.3b	35	20,27,31,*35*,39,42,48,49,66	4
8.3c	32	24,27,32,*32*,38,42,43,45,59	4
8.3d	39	22,27,32,36,37,*39*,49,53,68	6

Figure 8.3(a) is the original image and Figures 8.3(b)–8.3(d) are the compressed images at the compression rates of 80%, 50%, and 20%, respectively. Let us randomly select a pixel p and a 3×3 neighborhood as shown in the bold box. The value of p (i.e., the central pixel) is originally ranked 5 and is now ranked 4 or 6 after compression, as shown in Table 8.1. Using the pixel-based features extraction algorithm and using $T_1 = 0.25$ and $T_2 = 0.75$, the pixel p will all receive 0 in the original and compressed images.

Rotation is one of the techniques that attacks watermarking. In order to extract pixel-based features accurately and avoid rotational attack, we design the *grayscale morphological structuring element* (GMSE) to be circular. Figure 8.4 shows circular GMSEs of different sizes. Apparently, the larger the GMSE is, the more circular it is. Note that, the hyphen symbol (–) indicates *don't care*, which is equivalent to placing $-\infty$ in the GMSE.

(a)

-	-	1	1	-	-
-	1	2	2	1	-
1	2	1	1	3	1
-	2	1	2	3	-
-	-	1	2	-	-

(b)

-	-	-	1	1	1	-	-	-
-	-	1	2	2	2	1	-	-
-	1	2	1	3	1	2	1	-
1	2	2	2	3	3	3	2	1
1	2	3	2	4	3	3	2	1
1	2	2	3	3	3	3	2	1
-	1	2	2	3	2	2	1	-
-	-	1	2	2	2	1	-	-
-	-	-	1	1	1	-	-	-

(c)

-	-	-	-	-	1	1	1	-	-	-	-	-
-	-	-	2	2	2	2	2	2	2	-	-	-
-	-	3	3	3	3	3	3	3	3	3	-	-
-	2	3	4	4	4	4	4	4	4	3	2	-
-	2	3	4	5	5	5	5	5	4	3	2	-
1	2	3	4	5	6	6	6	5	4	3	2	1
1	2	3	4	5	6	7	6	5	4	3	2	1
1	2	3	4	5	6	6	6	5	4	3	2	1
-	2	3	4	5	5	5	5	5	4	3	2	-
-	2	3	4	4	4	4	4	4	4	3	2	-
-	-	3	3	3	3	3	3	3	3	3	-	-
-	-	-	2	2	2	2	2	2	2	-	-	-
-	-	-	-	-	1	1	1	-	-	-	-	-

FIGURE 8.4 Circular GMSEs.

8.3 THE STRATEGIES FOR ADJUSTING THE VARYING-SIZED TRANSFORM WINDOW AND QUALITY FACTOR

The size of the VSTW can determine whether the spatial or frequency domain is employed. For example, if the size is 1×1, it is equivalent to the spatial domain approach. If the size is 2×2, 4×4, or 8×8, we split the original image using the size of subblocks accordingly, and perform DCT on each subblock, and it belongs to the frequency domain approach. Note that the size of the VSTW could be arbitrary, not necessarily a square of a power of two.

In principle, fragile watermarks are embedded into LSBs, but robust watermarks are into bits of higher significance. In order to integrate both watermarks, the QFs are developed based on the spread spectrum. We know that shifting the binary bits of x to left (denoted as *shl*) or right (denoted as *shr*) by y bits is equivalent to multiplying or dividing x by 2^y. The position for embedding the watermark can be determined based on the QF, as shown in Table 8.2. For example, if the watermark is embedded into the 4th bit (i.e., QF = 8) of a pixel, we will divide the pixel value by 8, and then replace the resulting LSB by the watermark. Finally, we multiply the result by 8 to obtain the watermarked value.

Cox et al. have proposed spread spectrum watermarking and indicated that watermarks should not be embedded into insignificant regions of the image or its spectrum since many signal and geometric processes can affect these components [1]. For robustness, they suggested that the watermarks should be embedded into the regions with large magnitudes in coefficients of a transformed image. For example, there are a few large magnitudes in the coefficients, as shown in Figure 8.5. Figure 8.5(a) shows the original image of size 16×16 in which all pixels are 210. Figure 8.5(b) is the transformed image by DCT using the 8×8 VSTW. There are four significant regions with the large magnitude. Figure 8.5(c) shows the results after embedding the watermarks into the 6th bit of four different locations as marked in the bold box. Note that the watermarks are not embedded into the four significant regions. After performing the IDCT on Figure 8.5(c), we obtain the watermarked image, as shown in Figure 8.5(d). It is obvious that the original image will have a

---

TABLE 8.2

The Relationship between the Quality Factor and the Embedded Position of the Watermark

Intensity	Binary	Embedded Watermark	Embedded Position	Quality Factor	After Embedding	New Intensity
41	0010100*1*	0	1 (LSB)	1	0010100*0*	40
43	0010*1*011	0	4	8	0010*0*011	35
66	010000*1*0	0	2	2	0100000*0*	64
110	0110*1*110	0	4	8	0110*0*110	102
210	110*1*0010	0	5	16	110*0*0010	194

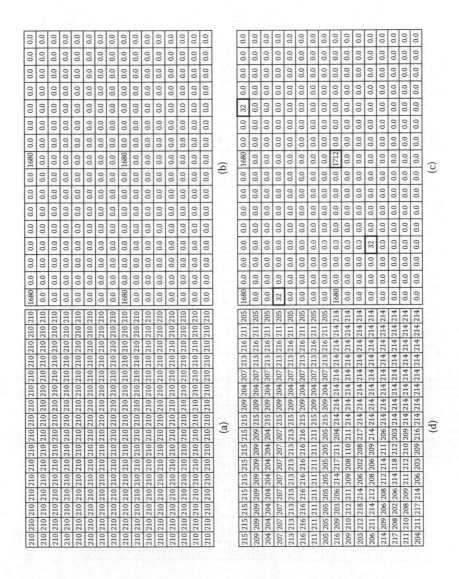

FIGURE 8.5 An example of embedding watermarks into different spectral regions.

FIGURE 8.6 An example of enlarging the capacity of a watermark.

TABLE 8.3

General Rules for Determining the Varying-Sized Transform Window and Quality Factor

Purpose	VSTW	QF
Robust, Spatial Domain	1×1	Larger than 1
Robust, Frequency Domain	Larger than 1×1	Larger than 1
Fragile, Spatial Domain	1×1	Less than or equal to 1
Fragile, Frequency Domain	Larger than 1×1	Less than or Equal to 1

huge degradation because the watermarks are embedded into nonsignificant regions. Therefore, the size of embedded watermarks is quite limited.

Figure 8.6 shows an example of our AP watermarking for enlarging the capacity of the watermarks. In order to achieve this, a small VSTW is selected. The watermark used is the 64 bits of all 1's. Considering robustness, the watermark is embedded into a bit of higher significance (e.g., the 5th bit). Given the size of the VSTW being 8×8, we obtain Figure 8.6(a) by embedding the watermark into Figure 8.5(b). After performing IDCT on Figure 8.6(a), a watermarked image is obtained, as shown in Figure 8.6(b), that is largely distorted. It is obvious that the performance is not good since too much data is embedded. However, if the 2×2 VSTW is selected, we can obtain Figure 8.6(c), which is the result after embedding the watermark into the frequency domain of the original image. Similarly, after performing IDCT on Figure 8.6(c), we can obtain the watermarked image as shown in Figure 8.6(d). Therefore, through the VSTW we can determine the size of each subimage to enlarge the capacity of the watermarks.

By adjusting the VSTW and the QF, our technique can be equivalent to existing watermarking techniques. For example, by setting the VSTW to be $N \times N$ and each element in the QF to be a large number, our AP watermarking is the same as the spread spectrum watermarking developed by Cox et al. [1]. To simulate the block-based fragile watermarking of Wong [2], we set the VSTW to be 1×1 and each element in the QF to be 1. The general rules for determining the VSTW and the QF are shown in Table 8.3.

8.4 EXPERIMENTAL RESULTS

Figures 8.7(a) and 8.7(b) show an original image and its pixel-based features, respectively. Figures 8.7(c) and (d) show the image after JPEG compression with a 20% quality level and its pixel-based features, respectively. By comparing Figures 8.7(b) and 8.7(d), we see that the pixel-based features remain almost the same even after low-quality JPEG compression. Table 8.4 provides the quantitative measures. Note that the size of the structuring element is 3×3 and (T_1, T_2) indicates the threshold values adopted in step 4 of the pixel-based features extraction algorithm. Table 8.5 provides the analysis based on different sizes of the structuring element.

(a)   (b)

(c)   (d)

FIGURE 8.7 An example of the pixel-based features of an image.

Figures 8.8(a) and 8.8(b) show an image and its data of size 16 × 16. The size of the VSTW adopted is 4 × 4 in this example. Figure 8.8(c) shows the QF corresponding to each subimage. Figure 8.8(f) is the original watermark. Note that the watermark is embedded into the direct current (DC) component of coefficients in

TABLE 8.4

Quantitative Measures for the Pixel-Based Features of an Image after Differing Qualities of JPEG Compression

JPEG Quality	PSNR	$T_1 = 0.1$ $T_2 = 0.9$	$T_1 = 0.15$ $T_2 = 0.85$	$T_1 = 0.2$ $T_2 = 0.8$	$T_1 = 0.25$ $T_2 = 0.75$	$T_1 = 0.3$ $T_2 = 0.7$	$T_1 = 0.35$ $T_2 = 0.65$
90 %	48.98	98.67%	98.31%	94.14%	98.05%	96.78%	95.73%
80 %	44.92	96.44%	95.98%	95.78%	95.67%	95.04%	93.21%
70 %	41.99	95.88%	95.09%	94.84%	94.65%	93.77%	91.84%
60 %	39.18	95.45%	94.91%	94.69%	94.20%	93.50%	91.31%
50 %	38.68	94.64%	94.04%	93.67%	93.37%	92.33%	90.20%
40 %	35.81	94.15%	93.48%	93.08%	92.49%	92.12%	90.06%
30 %	32.33	92.22%	91.24%	90.73%	90.42%	89.62%	86.44%
20 %	30.57	91.45%	90.38%	89.93%	89.38%	88.47%	86.78%
10 %	24.51	74.11%	74.08%	73.96%	74.15%	75.09%	78.73%

TABLE 8.5

Pixel-Based Features Using Different Sizes of the Structuring Elements

JPEG Quality	SE = 5 × 5		SE = 7 × 7	
	$T_1 = 0.1\ T_2 = 0.9$	$T_1 = 0.25\ T_2 = 0.75$	$T_1 = 0.1\ T_2 = 0.9$	$T_1 = 0.25\ T_2 = 0.75$
90 %	98.75%	98.27%	99.10%	98.68%
80 %	98.15%	98.27%	98.56%	98.98%
70 %	98.38%	96.05%	98.71%	96.67%
60 %	98.16%	95.87%	96.65%	96.35%
50 %	96.30%	94.60%	96.52%	95.55%
40 %	95.92%	94.43%	96.38%	95.06%
30 %	94.32%	91.73%	95.04%	93.18%
20 %	93.90%	91.76%	94.37%	92.83%
10 %	76.89%	76.78%	79.30%	79.18%

the transformed subimages. That is, each subimage is given 1 bit of watermark. Figures 8.8(d) and 8.8(e) show the watermarked image and its data. The JPEG compression is adopted as the attacking method. Pairs of Figures 8.8(g) and 8.8(h), 8.8(j) and 8.8(k), 8.8(m) and 8.8(n), and 8.8(p) and 8.8(q) are the results of JPEG compression at the quality levels 80%, 60%, 40%, and 20%, respectively. Figures 8.8(i), 8.8l, 8.8(o), and 8.8(r) are the corresponding extracted watermarks.

Figure 8.9 shows an example of our AP watermarking algorithm for extracting the embedded watermarks after different JPEG compression qualities. Figures 8.9(a)1, 8.9(b)1 and 8.9(c)1 are the original watermarks of sizes 128×128, 64×64, and 32×32, respectively. Sizes of the VSTWs used for embedding those watermarks are 2×2, 4×4, and 8×8, and their corresponding QFs are shown in Figures 8.10(a)–8.10(c). After embedding watermarks into the host image, we use four JPEG compression qualities—80%, 60%, 40%, and 20%—as the attackers. Figures 8.9(a)2, 8.9(b)2, and 8.9(c)2 are the extracted watermarks after 80% JPEG compression. Similarly, Figures 8.9(a)3, 8.9(b)3, and 8.9(c)3; 8.9(a)4, 8.9(b)4, and 8.9(c)4; and 8.9(a)5, 8.9(b)5, and 8.9(c)5 are obtained after 60%, 40%, and 20% JPEG compressions, respectively. Figures 8.8 and 8.9 demonstrate that our AP watermarking can resist a JPEG compression attack.

8.5 THE COLLECTING APPROACH FOR GENERATING THE VSTW

In order to extend our AP watermarking to general images whose size is not quadrate, we develop a collecting approach for generating the VSTW. Unlike the traditional approach to separating an original image into many nonoverlapping blocks, the collecting approach systematically picks up proper pixels to form the the VSTW. Figure 8.11 illustrates the collecting approach. Figure 8.11(a) is the original image. We collect 4, 16, and 64 pixels to generate VSTWs of size 2×2, 4×4, and 8×8,

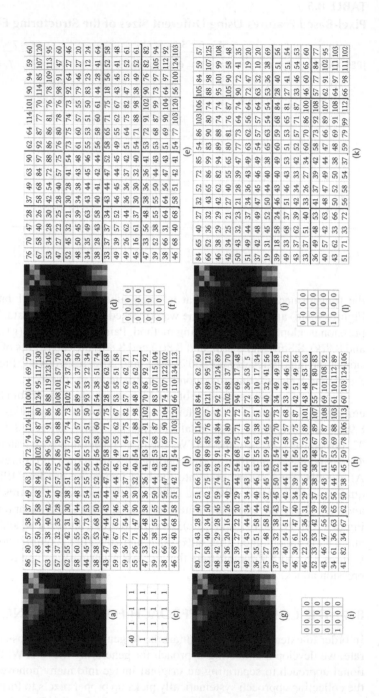

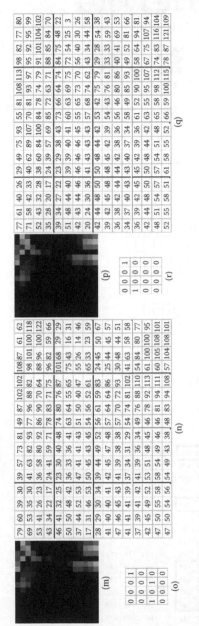

FIGURE 8.8 An example of our AP watermarking technique.

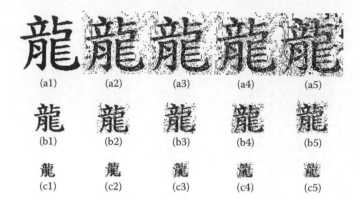

(a1)   (a2)   (a3)   (a4)   (a5)

(b1)   (b2)   (b3)   (b4)   (b5)

(c1)   (c2)   (c3)   (c4)   (c5)

FIGURE 8.9 An example of extracting embedded watermarks after JPEG compression.

16	1	1	1	1	1	1	1
1	1	1	1	1	1	1	1
1	1	1	1	1	1	1	1
1	1	1	1	1	1	1	1
1	1	1	1	1	1	1	1
1	1	1	1	1	1	1	1
1	1	1	1	1	1	1	1
1	1	1	1	1	1	1	1

16	1	1	1
1	1	1	1
1	1	1	1
1	1	1	1

(b)

16	1
1	1

(c)

(a)

FIGURE 8.10 The QFs for their corresponding VSTWs.

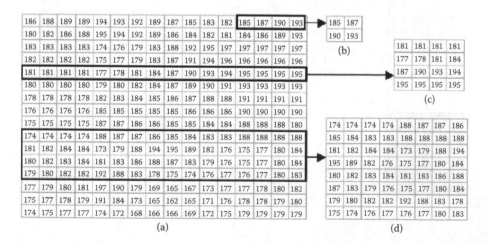

FIGURE 8.11 An example of the collecting approach for generating the VSTW.

as shown in Figures 8.11(b)–8.11(d). Due to the space correlation, most of coefficients of transformed blocks obtained by traditional approach are 0 except the low-frequency parts. Our collecting approach will break the space correlation. Therefore, it not only extends our AP watermarking to general-sized images, but also enlarges the capacity of watermarks when the frequency-domain approach is used.

REFERENCES

[1] Cox, I., et al. "Secure Spread Spectrum Watermarking for Multimedia." *IEEE Trans. Image Processing* 6 (1997): 1673.

[2] Wong, P. W. "A Public Key Watermark for Image Verification and Authentication." In *Proc. IEEE Int. Conf. Image Processing.* Chicago, 1998.

[3] Rivest, R., A. Shamir, and L. Adleman. "A Method for Obtaining Digital Signatures and Public-Key Cryptosystems." *Communications of the ACM* 21 (1978): 120.

[4] Rivest, R. L.. "The MD5 Message Digest Algorithm." RFC 1321, 1992. Available online at http://www.faqs.org/rfcs/rfc1321.html.

[5] Celik, M. U., et al. "Hierarchical Watermarking for Secure Image Authentication with Localization." *IEEE Trans. Image Processing* 11 (2002): 585.

[6] Chen, L.-H., and J.-J. Lin. "Mean Quantization Based Image Watermarking." *Image Vision Computing* 21 (2003): 717.

[7] Serra, J. *Image Analysis and Mathematical Morphology.* New York: Academic Press, 1982.

[8] Haralick, R. M., S. R. Sternberg, and X. Zhuang. "Image Analysis Using Mathematical Morphology." *IEEE Trans. Pattern Anal. Mach. Intell.* 9 (1988): 532.

[9] Shih, F. Y. and O. R. Mitchell. "Threshold Decomposition of Grayscale Morphology into Binary Morphology, *IEEE Trans. Pattern Anal. Mach. Intell.* 11 (1989): 31.

9 Robust High-Capacity Digital Watermarking

Since multimedia technologies have become increasingly sophisticated in the rapidly growing field of Internet applications, data security, including copyright protection and data integrity detection, has become a tremendous concern. One solution for achieving data security is digital watermarking technology, which embeds hidden information or secret data in a host image [1–3]. This serves as a suitable tool for identifying the source, creator, owner, distributor, or authorized consumer of a document or image. It can also be used to detect whether a document or image is illegally distributed or modified.

There are two domain-based watermarking techniques: one in the spatial domain and the other in the frequency domain. In the spatial domain [4–8], we embed watermarks into a host image by changing the gray levels of certain pixels. The embedding capacity may be large, but the hidden information could be easily detected by means of computer analysis. In the frequency domain [9–17], we insert watermarks into frequency coefficients of the image transformed by discrete cosine transform (DCT), discrete fourier transform (DFT), or discrete wavelet transform (DWT). The hidden information is in general difficult to detect, but we cannot embed a large volume of watermarking in the frequency domain due to image distortion.

It is reasonable to think that frequency-domain watermarking would be robust since the embedded watermark is spread out all over the spatial extent of an image [9]. If the watermark is embedded into locations of large absolute values (known as *significant coefficients*) of the transformed image, the watermarking technique will become more robust. Unfortunately, the transformed images in general contain only a few significant coefficients, so the watermarking capacity is limited.

9.1 THE WEAKNESS OF CURRENT ROBUST WATERMARKING

There are several approaches for achieving robust watermarking such that the watermarks are detectable after distorting the watermarked images. However, the capacity of robust watermarking techniques is usually limited due to their strategy. For example, the redundant embedding approach achieves robustness by embedding more than one copy of the same watermark into an image; however, the multiple copies reduce the size of the individual watermarks. For significant-coefficients embedding, it is obvious that the capacity for watermarks is due to the number of significant coefficients. Unfortunately, this number is quite limited in most images due to local spatial similarity. For region-based embedding, we embed a bit of watermark over the region of an image in order to spread out the message. Hence, the capacity of watermarks is restricted to block size.

9.2 THE CONCEPT OF ROBUST WATERMARKING

Zhao, Chen, and Liu have presented a robust wavelet-domain watermarking algorithm based on the chaotic map [10]. They divide an image into a set of 8×8 blocks with labels. After the order is mixed via a chaotic map, the first 256 blocks are selected for embedding watermarks. Miller, Doerr, and Cox [11] have proposed a robust high-capacity (RHC) watermarking algorithm through informed coding and embedding. However, they can embed only 1,380 bits of information in an image of size 240×368—that is, a capacity of 0.015625 bits/pixel.

There are two criteria to be considered when we develop the RHC watermarking technique. First, the strategy for embedding and extracting watermarks ensures robustness. Second, the strategy for enlarging the capacity does not affect the robustness of the watermarking. Frequency-domain watermarking possesses strong robustness since the embedded messages are spread out all over the spatial extent of an image [9]. Moreover, if the messages are embedded into significant coefficients, the watermarking technique will be more robust. Therefore, the significant-coefficients embedding approach is adopted as the basis of RHC watermarking to satisfy the first criterion. The remaining problem is to increase capacity without degrading robustness.

9.3 ENLARGEMENT OF SIGNIFICANT COEFFICIENTS

During spatial-frequency transformation, the low frequencies in the transformed domain reflect smooth areas of an image, and the high frequencies reflect the areas with large intensity changes, such as edges and noise. Therefore, due to local spatial similarity the significant coefficients of a transformed image are limited. Unless an image contains heavy noise levels, we cannot obtain large significant coefficients through any of the transformation approaches (DCT, DFT, or DWT).

9.3.1 Breaking the Local Spatial Similarity

The noise in an image leads us to enlarge the capacity. If we could rearrange pixel positions such that the rearranged image contains lots of noise, the significant coefficients of the rearranged image would be increased dramatically. Therefore, we adopt the chaotic map to relocate pixels.

Figure 9.1 illustrates an example of increasing the number of significant coefficients by rearranging pixel locations. Figure 9.1(a) shows an image; Figure 9.1(b) shows the same image with its gray values displayed. Figure 9.1(d) shows its relocated image, with its gray values displayed in 9.1(e). Figures 9.1(c) and 9.1(f) are obtained by applying DCT on Figures 9.1(b) and 9.1(e), respectively. If the threshold is set to be 70, we obtain 14 significant coefficients in Figure 9.1(f), but only 2 significant coefficients in 9.1(c).

9.3.2 The Block-Based Chaotic Map

In order to enlarge the watermark capacity, we adopt the chaotic map to break the local spatial similarity of an image in order to generate more significant coefficients

(a)

(b)

215	201	177	145	111	79	55	41
215	201	177	145	111	79	55	41
215	201	177	145	111	79	55	41
215	201	177	145	111	79	55	41
215	201	177	145	111	79	55	41
215	201	177	145	111	79	55	41
215	201	177	145	111	79	55	41
215	201	177	145	111	79	55	41

(c)

1024.0	499.4	0.0	1.7	0.0	1.2	0.0	1.4
0.0	0.0	0.0	0.0	0.0	0.0	0.0	0.0
0.0	0.0	0.0	0.0	0.0	0.0	0.0	0.0
0.0	0.0	0.0	0.0	0.0	0.0	0.0	0.0
0.0	0.0	0.0	0.0	0.0	0.0	0.0	0.0
0.0	0.0	0.0	0.0	0.0	0.0	0.0	0.0
0.0	0.0	0.0	0.0	0.0	0.0	0.0	0.0
0.0	0.0	0.0	0.0	0.0	0.0	0.0	0.0

(d)

(e)

215	201	177	145	111	79	55	41
201	177	145	111	79	55	41	215
177	145	111	79	55	41	215	201
145	111	79	55	41	215	201	177
111	79	55	41	215	201	177	145
79	55	41	215	201	177	145	111
55	41	215	201	177	145	111	79
41	215	201	177	145	111	79	55

(f)

1024.0	0.0	0.0	0.0	0.0	0.0	0.0	0.0
0.0	110.8	244.0	-10.0	35.3	-2.6	10.5	-0.5
0.0	244.0	-116.7	-138.2	0.0	-27.5	0.0	-6.4
0.0	-10.0	-138.2	99.0	72.2	0.2	13.3	0.3
0.0	35.3	0.0	72.2	-94.0	-48.2	0.0	-7.0
0.0	-2.6	-27.5	0.2	-48.2	92.3	29.9	0.4
0.0	10.5	0.0	13.3	0.0	29.9	-91.3	-15.7
0.0	-0.5	-6.4	0.3	-7.0	0.4	-15.7	91.9

FIGURE 9.1 An example of increasing the number of significant coefficients: (a) and (b) an image and its gray values; (c) the DCT coefficients; (d) and (e) the relocated image and its gray values, (f) the DCT coefficients.

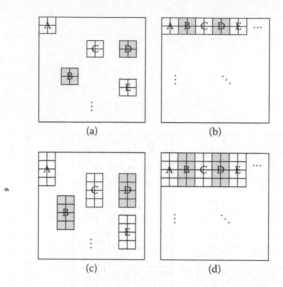

(a)   (b)

(c)   (d)

FIGURE 9.2 An example of the block-based relocation: (a) and (b) a diagram and its relocated result based on a 2×2 block; (c) and (d) a diagram and its relocated result based on a 2×4 block.

in the transformed domain. Since the traditional chaotic map is pixel-based and not suited to generate the reference register [6, 10, 16], we develop a new block-based chaotic map to break the local spatial similarity. In other words, the block-based chaotic map for relocating pixels is based on the block unit (i.e., a set of connected pixels) instead of the pixel unit.

Figures 9.2(a) and 9.2(b) show a diagram and its relocated result based on a block size of 2×2. Figures 9.2(c) and 9.2(d) show a diagram and its relocated result based on a block size of 2×4. In general, the bigger the block size is, the larger the local similarity is. We apply the block size of 2×2 and $l = 2$ on Figure 9.3, where 9.3(a) is the Lena image, 9.3(b) its relocated image, and 9.3(c) the resulting image in the next iteration.

(a)   (b)   (c)

FIGURE 9.3 An example of performing the block-based chaotic map: (a) a Lena image; (b) the relocated image; (c) the continuously relocated image.

9.4 THE DETERMINATION OF EMBEDDING LOCATIONS

9.4.1 INTERSECTION-BASED PIXELS COLLECTION

Intersection-based pixels collection (IBPC) labels an image using two symbols alternatively and then collecting the two subimages with the same symbol. Figures 9.4(a)–9.4(c) show three different approaches to collecting the pixels with the same label horizontally, vertically, and diagonally, respectively. The two subimages formed are shown in Figures 9.4(d) and 9.4(e). Note that the pair of images obtained has local spatial similarity, even after transformation or attacks. From experiments, the diagonal IBPC shows better similarity in our RHC watermarking algorithm. Therefore, we generate a pair of 8×8 coefficients from the 16×8 image by using the IBPC approach followed by DCT. Afterward, one is used as the reference register for indicating significant DCT coefficients and the other is used as the container for embedding watermarks.

9.4.2 THE REFERENCE REGISTER AND CONTAINER

An example of demonstrating the similarity of two subimages obtained by IBPC is shown in Figure 9.5. Figure 9.5(a) shows the Lena image; 9.5(b) shows the same image with a small block of pixels cropped. Figures 9.5(c) and 9.5(d) show the pair of subimages obtained from 9.5(b) via IBPC. Figures 9.5(e) and 9.5(f) are their respective DCT images. Figures 9.5(g) and 9.5(h) show the results after dividing 9.5(e) and 9.5(f), respectively, by the quantization table with a quality factor (QF) of 50. Note that the QF will be used in equation 9.1, and an example of the quantization table in JPEG compression is given in Figure 9.6. We observe that Figures 9.5(g) and 9.5(h) are similar. Therefore, either one can be used as the reference register to indicate the significant coefficients for embedding watermarks. The significant coefficients have the values larger than a predefined threshold RR_{Th}.

```
(a)
A A A A A A A A A A A A A A A A
B B B B B B B B B B B B B B B B
A A A A A A A A A A A A A A A A
B B B B B B B B B B B B B B B B
A A A A A A A A A A A A A A A A
B B B B B B B B B B B B B B B B
A A A A A A A A A A A A A A A A
B B B B B B B B B B B B B B B B

(b)
A B A B A B A B A B A B A B A B
A B A B A B A B A B A B A B A B
A B A B A B A B A B A B A B A B
A B A B A B A B A B A B A B A B
A B A B A B A B A B A B A B A B
A B A B A B A B A B A B A B A B
A B A B A B A B A B A B A B A B
A B A B A B A B A B A B A B A B
```

```
(c)
A B A B A B A B A B A B A B A B
B A B A B A B A B A B A B A B A
A B A B A B A B A B A B A B A B
B A B A B A B A B A B A B A B A
A B A B A B A B A B A B A B A B
B A B A B A B A B A B A B A B A
A B A B A B A B A B A B A B A B
B A B A B A B A B A B A B A B A

(d)               (e)
A A A A A A A A   B B B B B B B B
A A A A A A A A   B B B B B B B B
A A A A A A A A   B B B B B B B B
A A A A A A A A   B B B B B B B B
A A A A A A A A   B B B B B B B B
A A A A A A A A   B B B B B B B B
A A A A A A A A   B B B B B B B B
A A A A A A A A   B B B B B B B B
```

FIGURE 9.4 An example of IBPC.

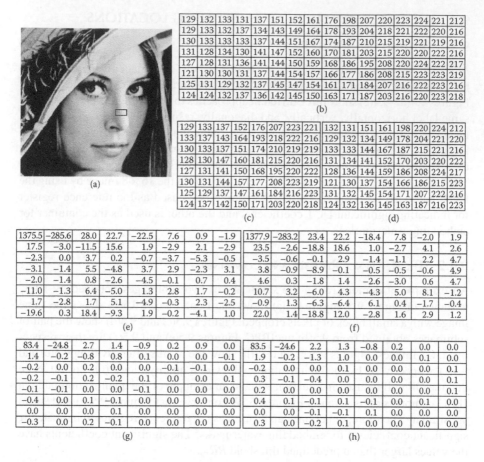

(b)

129	132	133	131	137	151	152	161	176	198	207	220	223	224	221	212
129	133	132	137	134	143	149	164	178	193	204	218	221	222	220	216
130	133	133	133	137	144	151	167	174	187	210	215	219	221	219	216
131	128	134	130	141	147	152	160	170	181	203	215	220	220	222	216
127	128	131	136	141	144	150	159	168	186	195	208	220	224	222	217
121	130	130	131	137	144	154	157	166	177	186	208	215	223	223	219
125	131	129	132	137	145	147	154	161	171	184	207	216	222	223	216
124	124	132	137	136	142	145	150	163	171	187	203	216	220	223	218

(c)

129	133	137	152	176	207	223	221
133	137	143	164	193	218	222	216
130	133	137	151	174	210	219	219
128	130	147	160	181	215	220	216
127	131	141	150	168	195	220	222
130	131	144	157	177	208	223	219
125	129	137	147	161	184	216	223
124	137	142	150	171	203	220	218

(d)

132	131	151	161	198	220	224	212
129	132	134	149	178	204	221	220
133	133	144	167	187	215	221	216
131	134	141	152	170	203	220	222
128	136	144	159	186	208	224	217
121	130	137	154	166	186	215	223
131	132	145	154	171	207	222	216
124	132	136	145	163	187	216	223

(e)

1375.5	-285.6	28.0	22.7	-22.5	7.6	0.9	-1.9
17.5	-3.0	-11.5	15.6	1.9	-2.9	2.1	-2.9
-2.3	0.0	3.7	0.2	-0.7	-3.7	-5.3	-0.5
-3.1	-1.4	5.5	-4.8	3.7	2.9	-2.3	3.1
-2.0	-1.4	0.8	-2.6	-4.5	-0.1	0.7	0.4
-11.0	-1.3	6.4	-5.0	1.3	2.8	1.7	-0.2
1.7	-2.8	1.7	5.1	-4.9	-0.3	2.3	-2.5
-19.6	0.3	18.4	-9.3	1.9	-0.2	-4.1	1.0

(f)

1377.9	-283.2	23.4	22.2	-18.4	7.8	-2.0	1.9
23.5	-2.6	-18.8	18.6	1.0	-2.7	4.1	2.6
-3.5	-0.6	-0.1	2.9	-1.4	-1.1	2.2	4.7
3.8	-0.9	-8.9	-0.1	-0.5	-0.5	-0.6	4.9
4.6	0.3	-1.8	1.4	-2.6	-3.0	0.6	4.7
10.7	3.2	-6.0	4.3	-4.3	5.0	8.1	-1.2
-0.9	1.3	-6.3	-6.4	6.1	0.4	-1.7	-0.4
22.0	1.4	-18.8	12.0	-2.8	1.6	2.9	1.2

(g)

83.4	-24.8	2.7	1.4	-0.9	0.2	0.9	0.0
1.4	-0.2	-0.8	0.8	0.1	0.0	0.0	-0.1
-0.2	0.0	0.2	0.0	0.0	-0.1	-0.1	0.0
-0.2	-0.1	0.2	-0.2	0.1	0.0	0.0	0.1
-0.1	-0.1	0.0	0.0	-0.1	0.0	0.0	0.0
-0.4	0.0	0.1	-0.1	0.0	0.0	0.0	0.0
0.0	0.0	0.0	0.1	0.0	0.0	0.0	0.0
-0.3	0.0	0.2	-0.1	0.0	0.0	0.0	0.0

(h)

83.5	-24.6	2.2	1.3	-0.8	0.2	0.0	0.0
1.9	-0.2	-1.3	1.0	0.0	0.0	0.1	0.0
-0.2	0.0	0.0	0.1	0.0	0.0	0.0	0.1
0.3	-0.1	-0.4	0.0	0.0	0.0	0.0	0.1
0.2	0.0	0.0	0.0	0.0	0.0	0.0	0.1
0.4	0.1	-0.1	0.1	-0.1	0.0	0.1	0.0
0.0	0.0	-0.1	-0.1	0.1	0.0	0.0	0.0
0.3	0.0	-0.2	0.1	0.0	0.0	0.0	0.0

FIGURE 9.5 An example of illustrating the similarity of two subimages obtained via IBPC: (a) A Lena image; (b) the extracted 16×8 image; (c) and (d) a pair of extracted 8×8 subimages; (e) and (f) the 8×8 DCT coefficients; (g) and (h) the quantized images.

16	11	10	16	24	40	51	61
12	12	14	19	26	58	60	55
14	13	16	24	40	57	69	56
14	17	22	29	51	87	80	62
18	22	37	56	68	109	103	77
24	35	55	64	81	104	113	92
49	64	78	87	103	121	120	101
72	92	95	98	112	100	103	99

FIGURE 9.6 A quantization table in the JPEG.

If $Q\,Table(i, j)$ denotes the quantization table, where $0 \leq i, j \leq 7$, the new quantization table $New\,Table(i, j)$ is obtained by

$$NewTable(i, j) = \begin{cases} QTable(i, j) \times \dfrac{50}{QF} & \text{if } QF < 50 \\ QTable(i, j) \times (2 - 0.02 \times QF) & \text{otherwise} \end{cases} \quad (9.1)$$

9.5 THE RHC WATERMARKING ALGORITHM

Our RHC watermarking algorithm contains two main components: the block-based chaotic map (BBCM) and the reference register.

9.5.1 THE EMBEDDING PROCEDURE

In order to explain our watermark embedding procedure clearly, we must first introduce the following symbols:

H The gray-level host image of size $n \times m$.

$H_{(i,j)}^{16 \times 8}$ The (i, j)-th subimage of size 16×8 of H, where $1 \leq i \leq [n/16]$ and $1 \leq j \leq [m/8]$.

$HA_{(i,j)}^{8 \times 8}$ and $HB_{(i,j)}^{8 \times 8}$ A pair of images obtained from $H_{(i,j)}^{16 \times 8}$ by IBPC.

$DA_{(i,j)}^{8 \times 8}$ and $DB_{(i,j)}^{8 \times 8}$ The transformed images of applying DCT on $HA_{(i,j)}^{8 \times 8}$ and $HB_{(i,j)}^{8 \times 8}$, respectively.

$QA_{(i,j)}^{8 \times 8}$ and $QB_{(i,j)}^{8 \times 8}$ The resulting images of dividing $DA_{(i,j)}^{8 \times 8}$ and $DA_{(i,j)}^{8 \times 8}$ by the quantization table, respectively.

$E_{(i,j)}^{8 \times 8}$ The watermarked image of $DA_{(i,j)}^{8 \times 8}$ using $QA_{(i,j)}^{8 \times 8}$ and $QB_{(i,j)}^{8 \times 8}$.

$M_{(i,j)}^{8 \times 8}$ The image after multiplying $E_{(i,j)}^{8 \times 8}$ by the quantization table.

$I_{(i,j)}^{8 \times 8}$ The image obtained by the inverse discrete cosine transformation (IDCT) of $M_{(i,j)}^{8 \times 8}$.

$C_{(i,j)}^{16 \times 8}$ The watermarked subimage obtained by combining $I_{(i,j)}^{8 \times 8}$ and $HB_{(i,j)}^{8 \times 8}$ using IBPC.

O The output watermarked image obtained by collecting all the subimages of $C_{(i,j)}^{16 \times 8}$.

We present the overall embedding procedure as follows, and in the flowchart in Figure 9.7.

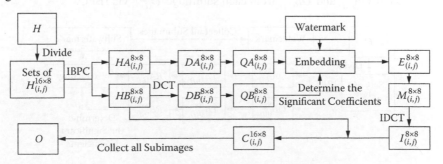

FIGURE 9.7 The embedding procedure for our RHC watermarking.

THE EMBEDDING PROCEDURE FOR OUR RHC WATERMARKING ALGORITHM

1. Divide an input image H into a set of subimages, $H_{(i,j)}^{16\times8}$, of size 16×9.
2. Build $HA_{(i,j)}^{8\times8}$ and $HB_{(i,j)}^{8\times8}$ from each subimage $H_{(i,j)}^{16\times8}$ via IBPC.
3. Obtain $DA_{(i,j)}^{8\times8}$ and $DB_{(i,j)}^{8\times8}$ from $HA_{(i,j)}^{8\times8}$ and $HB_{(i,j)}^{8\times8}$ via DCT.
4. Obtain $QA_{(i,j)}^{8\times8}$ and $QB_{(i,j)}^{8\times8}$ by dividing $DA_{(i,j)}^{8\times8}$ and $DB_{(i,j)}^{8\times8}$ by the JPEG quantization table.
5. Determine the proper positions of significant coefficients in $QB_{(i,j)}^{8\times8}$ and embed watermarks into the corresponding positions in $QA_{(i,j)}^{8\times8}$ to obtain $E_{(i,j)}^{8\times8}$. (The detailed embedding strategy will be described in section 9.5.3.1.)
6. Obtain $M_{(i,j)}^{8\times8}$ by multiplying $E_{(i,j)}^{8\times8}$ by the JPEG quantization table.
7. Obtain $I_{(i,j)}^{8\times8}$ by applying IDCT on $M_{(i,j)}^{8\times8}$.
8. Reconstruct $C_{(i,j)}^{16\times8}$ by combining $I_{(i,j)}^{8\times8}$ and $HB_{(i,j)}^{8\times8}$.
9. Obtain the output watermarked image Q by collecting all the $C_{(i,j)}^{16\times8}$'s.

9.5.2 THE EXTRACTING PROCEDURE

After receiving the watermarked image O, we intend to extract the watermark information. The following symbols are introduced in order to explain the watermark extracting procedure:

$O_{(i,j)}^{16\times8}$ The (i, j)th subimage of size 16×8 of O.

$OA_{(i,j)}^{8\times8}$ and $OB_{(i,j)}^{8\times8}$ A pair of images extracted from $O_{(i,j)}^{16\times8}$ using IBPC.

$TA_{(i,j)}^{8\times8}$ and $TB_{(i,j)}^{8\times8}$ The transformed image after applying DCT on $OA_{(i,j)}^{8\times8}$ and $OB_{(i,j)}^{8\times8}$.

$RA_{(i,j)}^{8\times8}$ and $RB_{(i,j)}^{8\times8}$ The resulting images after dividing $TA_{(i,j)}^{8\times8}$ and $TB_{(i,j)}^{8\times8}$ by the quantization table.

We present the watermark extracting procedure as follows, and in the flowchart in Figure 9.8.

THE EXTRACTING PROCEDURE FOR OUR RHC WATERMARKING ALGORITHM

1. Divide the watermarked image O into a set of subimages, $O_{(i,j)}^{16\times8}$, of size 16×9.
2. Build $OA_{(i,j)}^{8\times8}$ and $OB_{(i,j)}^{8\times8}$ from each subimage $O_{(i,j)}^{16\times8}$ via IBPC.

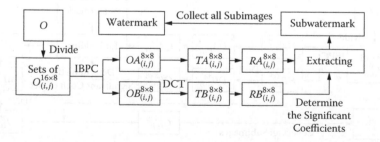

FIGURE 9.8 The extracting procedure for our RHC watermarking.

3. Obtain $TA_{(i,j)}^{8\times8}$ and $TB_{(i,j)}^{8\times8}$ from $OA_{(i,j)}^{8\times8}$ and $OB_{(i,j)}^{8\times8}$ via DCT.
4. Obtain $RA_{(i,j)}^{8\times8}$ and $RB_{(i,j)}^{8\times8}$ by dividing $TA_{(i,j)}^{8\times8}$ and $TB_{(i,j)}^{8\times8}$ by the JPEG quantization table.
5. Determine the proper positions of significant coefficients in $RB_{(i,j)}^{8\times8}$ and extract the subwatermark from the corresponding positions in $RA_{(i,j)}^{8\times8}$. (The detailed extracting strategy will be described in section 9.5.3.2.)
6. Collect all the subwatermarks to obtain the watermark.

9.5.3 THE EMBEDDING AND EXTRACTING STRATEGIES

9.5.3.1 The Embedding Strategy

This section will describe the embedding and extracting strategies of our RHC watermarking algorithm. For embedding, each pair of $QA_{(i,j)}^{8\times8}$ (container) and $QB_{(i,j)}^{8\times8}$ (reference register) is obtained first. After the significant coefficients are determined by the reference register, the watermarks are embedded into the corresponding positions of the cover coefficients by adding the values as in equation 9.2. If $V_{(k,l)}^{C}$ and $V_{(k,l)}^{R}$ denote the values of $QA_{(i,j)}^{8\times8}$ and $QB_{(i,j)}^{8\times8}$, respectively, $S_{(k,l)}^{C}$ is the result after embedding the message:

$$S_{(k,l)}^{C} = \begin{cases} V_{(k,l)}^{C} + \alpha V_{(k,l)}^{R} & \text{if } V_{(k,l)}^{R} \geq RR_{Th} \\ V_{(k,l)}^{C} & \text{otherwise} \end{cases} \qquad (9.2)$$

where $0 \leq k,l \leq 7$ and $\alpha > 0$. Note that the bigger α is, the higher robustness is. The 8×8 $S_{(k,l)}^{C}$, s are collected to form the corresponding $E_{(i,j)}^{8\times8}$. The embedding strategy is presented below.

THE WATERMARK EMBEDDING STRATEGY

1. Determine the embedding location by checking the significant coefficients of $QB_{(i,j)}^{8\times8}$.
2. If the embedding message is 1 and $V_{(k,l)}^{R} \geq RR_{Th}$, we obtain $S_{(k,l)}^{C} = V_{(k,l)}^{C} + \alpha V_{(k,l)}^{R}$; otherwise, we set $S_{(k,l)}^{C} = V_{(k,l)}^{C}$.

9.5.3.2 The Extracting Strategy

For the extracting strategy, each pair of $RA_{(i,j)}^{8\times8}$ (container) and $RB_{(i,j)}^{8\times8}$ (reference register) is obtained first. After the embedded positions of the reference register are determined, the watermark can be extracted via equation 9.3 and the watermarked coefficients can be used to construct the original coefficients by equation 9.4. If $W_{(k,l)}^{C}$ and $W_{(k,l)}^{R}$ denote the values of $RA_{(i,j)}^{8\times8}$ and $RB_{(i,j)}^{8\times8}$, respectively, then $F_{(k,l)}^{C}$ is the result after the additional amount is removed by equation 9.4, and w is the embedded message. We have

$$w = \begin{cases} 1 & \text{if } (W_{(k,l)}^{C} - W_{(k,l)}^{R}) \geq \dfrac{\alpha}{2} V_{(k,l)}^{R} \\ 0 & \text{elsewhere} \end{cases} \qquad (9.3)$$

$$F_{(k,l)}^{C} = \begin{cases} W_{(k,l)}^{C} - \alpha W_{(k,l)}^{R} & \text{if } w = 1 \\ W_{(k,l)}^{C} & \text{otherwise} \end{cases} \qquad (9.4)$$

The embedding strategy of our RHC watermarking is presented below.

THE WATERMARK EXTRACTING STRATEGY

1. Determine the extracting location by checking the significant coefficients of $RB_{(i,j)}^{8\times8}$.
2. Obtain the embedded message using equation 9.2.
3. If the embedded message is 1, we calculate $F_{(k,l)}^C = W_{(k,l)}^C - \alpha W_{(k,l)}^R$; otherwise, we set $F_{(k,l)}^C = W_{(k,l)}^C$.

9.6 EXPERIMENTAL RESULTS

This section, we will give an example to illustrate that our algorithm can enlarge the watermarking capacity by using the BBCM and IBPC approaches. We will provide the experimental results on 200 images and the comparisons with the *iciens* algorithm of Miller et al. [11].

9.6.1 CAPACITY ENLARGEMENT

We give an example of the embedding strategy in Figure 9.9. Figure 9.9(a) is the relocated image of Figure 9.3(a) after performing seven iterations of relocation using the BBCM in equation 9.1, where the block size is 2×2 and $l = 2$. After obtaining Figure 9.9(b) from the small box in 9.9(a), we generate Figures 9.9(c) and 9.9(d) via IBPC. Figures 9.9(e) and 9.9(f) are respectively obtained from Figures 9.9(c) and 9.9(d) via DCT, followed by a division of the quantization table. Note that Figure 9.9(e) is considered the container for watermark embedding, and Figure 9.9(f) is the reference register for indicating the embedding positions, as shaded in gray. Since the number of embedding positions is 8, we set the watermark to be 11111111. Figure 9.9(g) shows the result after embedding watermarks into these calculated positions by equation 9.2, where $\alpha = 0.5$ and $RR_{Th} = 9$.

Figure 9.10 shows an example of the extracting strategy. Figure 9.10(a) is the watermarked subimage, and Figures 9.10(b) and 9.10(c) are extracted from 9.10(a) via IBPC. Note that since Figure 9.10(c) is exactly the same as Figure 9.9(d), the reference register is the same as in Figure 9.9(f). Therefore, after obtaining Figure 9.10(d) from 9.9(b) via DCT, we can extract the watermark 11111111 from the coefficients and generate Figure 9.10(e). Figure 9.10(f) is the reconstructed result. We observe that Figure 9.10(f) is exactly the same as the original image in Figure 9.9(b).

9.6.2 ROBUST EXPERIMENTS

Miller et al. have presented a robust high-capacity watermarking by applying informed coding and embedding [11]. They conclude that the watermarking scheme is robust if the message error rate (MER) is lower than 20%. In order to evaluate the robustness of our watermarking algorithm, we test 200 images and attack the watermarked images with JPEG compression, Gaussian noise, and low-pass filtering. The resulting MERs from these attacked images are provided.

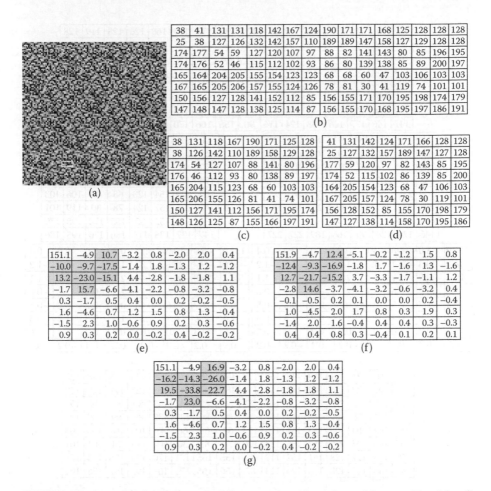

FIGURE 9.9 An example of the embedding strategy: (a) a relocated image; (b) a 16×8 subimage; (c) and (d) a pair of extracted results using IBPC; (e) a container; (f) a reference register; (g) the watermarked result.

Figure 9.11 shows the robustness experiment under JPEG compression. Figure 9.11(a) is the original Lena image. After continuously performing nine iterations of the BBCM, we obtain Figure 9.11(b), in which we embed watermarks by using the following parameters: a block size of 2×2, $l = 2$, $\alpha = 0.2$, and $RR_{Th} = 1$. Figure 9.11(c) is the watermarked image obtained by continuously performing eight iterations of the BBCM to Figure 9.11(b). Figure 9.11(d) is the attacked image after applying JPEG compression on Figure 9.11(c) using a QF of 20. Figure 9.11(e) is the reconstructed image after the embedded watermarks are removed. Note that our RHC watermarking not only is robust due to its MER of 0.19, which is less than 20%, but also contains the high capacity of 0.116 bits/pixel, which is much larger than the 0.015625 bits/pixel in Miller et al. [11].

(a)

11	41	116	131	121	142	184	124	211	171	184	168	126	128	119	128
25	12	127	108	132	136	157	115	189	199	147	168	127	134	128	129
160	177	40	59	114	120	97	97	82	82	140	143	84	85	203	195
174	188	52	49	115	102	102	73	86	58	139	122	85	82	200	197
200	164	225	205	155	154	104	123	40	68	35	47	86	106	93	103
167	203	205	232	157	162	124	115	78	61	30	23	119	63	101	95
169	156	140	128	145	152	110	85	153	155	173	170	205	198	190	179
147	146	127	126	138	124	114	90	156	166	170	189	195	227	186	227

(b)

11	116	121	184	211	184	126	119
12	108	136	115	199	168	134	129
160	40	114	97	82	140	84	203
188	49	102	73	58	122	82	197
200	225	155	104	40	35	86	93
203	232	162	115	61	23	63	95
169	140	145	110	153	173	205	190
146	126	124	90	166	189	227	227

(c)

41	131	142	124	171	168	128	128
25	127	132	157	189	147	127	128
177	59	120	97	82	143	85	195
174	52	115	102	86	139	85	200
164	205	154	123	68	47	106	103
167	205	157	124	78	30	119	101
156	128	152	85	155	170	198	179
147	127	138	114	156	170	195	186

(d)

151.0	-5.0	16.8	-3.1	0.8	-2.0	2.0	0.4
-16.2	-14.3	-26.1	-1.3	1.8	-1.3	1.2	-1.2
19.5	-33.9	-22.7	4.4	-2.7	-1.8	-1.8	1.1
-1.7	23.0	-6.6	-4.1	-2.3	-0.8	-3.2	-0.8
0.3	-1.7	0.5	0.4	0.1	0.2	-0.2	-0.5
1.6	-4.6	0.7	1.2	1.5	0.8	1.3	-0.3
-1.5	2.3	1.0	-0.6	0.9	0.2	0.3	-0.6
0.9	0.2	0.2	0.0	-0.2	0.4	-0.2	-0.2

(e)

151.0	-5.0	10.6	-3.1	0.8	-2.0	2.0	0.4
-10.0	-9.7	-17.6	-1.3	1.8	-1.3	1.2	-1.2
13.2	-23.0	-15.1	4.4	-2.7	-1.8	-1.8	1.1
-1.7	15.7	-6.6	-4.1	-2.3	-0.8	-3.2	-0.8
0.3	-1.7	0.5	0.4	0.1	0.2	-0.2	-0.5
1.6	-4.6	0.7	1.2	1.5	0.8	1.3	-0.3
-1.5	2.3	1.0	-0.6	0.9	0.2	0.3	-0.6
0.9	0.2	0.2	0.0	-0.2	0.4	-0.2	-0.2

(f)

38	41	131	131	118	142	167	124	190	171	171	168	125	128	128	128
25	38	127	126	132	142	157	110	189	189	147	158	127	129	128	128
174	177	54	59	127	120	107	97	88	82	141	143	80	85	196	195
174	176	52	46	115	112	102	93	86	80	139	138	85	89	200	197
165	164	204	205	155	154	123	123	68	68	60	47	103	106	103	103
167	165	205	206	157	155	124	126	78	81	30	41	119	74	101	101
150	156	127	128	141	152	112	85	156	155	171	170	195	198	174	179
147	148	127	128	138	125	114	87	156	155	170	168	195	197	186	191

FIGURE 9.10 An example of the extracting strategy: (a) a 16×8 watermarked image; (b) and (c) a pair of extracted 8×8 results via IBPC; (d) an 8×8 watermarked image via DCT; (e) the result after the embedded data are removed; (f) a reconstructed 16×8 image.

Figures 9.12 and 9.13 show the experimental results under low-pass filter and Gaussian noise attacks, respectively. The parameters used are the same as the previous example. In Figure 9.12, a 3×3 low-pass filter employing all 1's is utilized as the attacker. The capacity and the MER are 0.117 and 16%, respectively. In Figure 9.13, a Gaussian noise with the standard deviation (SD) 500 and mean 0 is added to the watermarked image. The capacity and the MER are 0.124 and 23%, respectively. It is obvious that our RHC watermarking algorithm not only achieves the robustness requirement but also maintains the high capacity of watermarks.

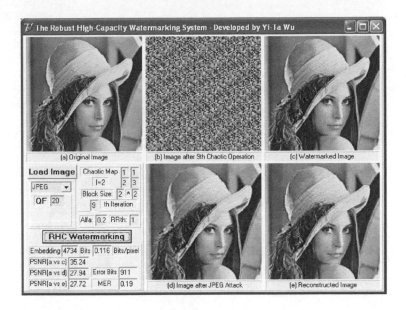

FIGURE 9.11 An example of the robustness experiment using JPEG compression: (a) a 202 × 202 Lena image; (b) the relocated image; (c) the watermarked image; (d) the image after JPEG compression; (e) the reconstructed image.

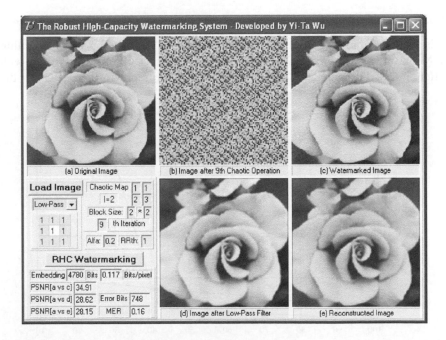

FIGURE 9.12 An example of the robustness experiment using a low-pass filter: (a) a 202 × 202 image; (b) the relocated image; (c) the watermarked image; (d) the image attacked by a low-pass filter; (e) the reconstructed image.

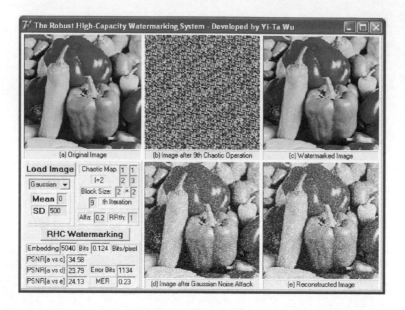

FIGURE 9.13 An example of robustness experiment using Gaussian noise: (a) a 202 × 202 image; (b) the relocated image; (c) the watermarked image; (d) the image attacked by Gaussian noise; (e) the reconstructed image.

9.6.3 PERFORMANCE COMPARISONS

This section compares our RHC watermarking algorithm with the iciens algorithm of Miller et al. [11]. Figure 9.14 shows the effect under JPEG compression with different QFs. For simplicity, we use an asterisk (*) to represent our RHC and the

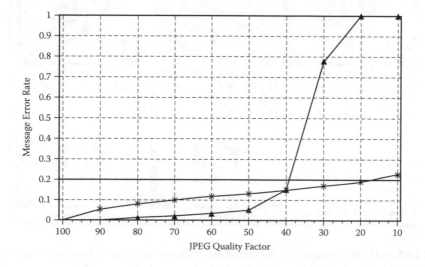

FIGURE 9.14 Robustness versus JPEG compression.

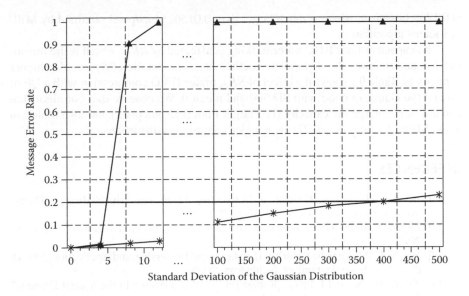

FIGURE 9.15 Robustness versus Gaussian noise.

symbol ▲ to represent iciens. Our algorithm has a slightly higher MER when the QF is higher than 50, but has a significantly lower MER when the QF is lower than 40.

Figure 9.15 shows the effect under Gaussian noise with mean 0 and different standard deviations. Our RHC algorithm outperforms the iciens algorithm in terms of low MER. Note that the parameters are a block size of 2×2, $l = 2$, $\alpha = 0.2$, and $RR_{Th} = 1$. For watermarking capacity, the average capacity of our RHC watermarking

TABLE 9.1
The Effect of RR_{Th}

RR_{Th}	Capacity (Bits)	Capacity (Bits/Pixel)	JPEG (QF = 20)		Gaussian Noise (SD = 500, Mean = 0)	
			Error Bits	MER	Error Bits	MER
10	810	0.0199	104	0.12	140	0.17
9	924	0.0226	120	0.13	170	0.18
8	1061	0.0260	139	0.13	195	0.18
7	1228	0.0301	163	0.13	228	0.19
6	1439	0.0353	197	0.14	271	0.19
5	1715	0.0420	243	0.14	329	0.19
4	2087	0.0511	304	0.15	403	0.19
3	2617	0.0641	404	0.15	522	0.20
2	3442	0.0844	577	0.18	712	0.21
1	5050	0.1238	968	0.19	1116	0.22

is 0.1238 bits/pixel, which is much larger than 0.015625 bits/pixel obtained by Miller's iciens algorithm.

The capacity of our RHC watermarking is affected by RR_{Th}, which is the threshold for determining the significant coefficients. The smaller RR_{Th} is, the higher capacity is. Table 9.1 shows the effect of RR_{Th} under JPEG compression with a QF of 20 and the Gaussian noise with SD 500 and mean 0. We observe that our algorithm can not only enlarge the capacity (i.e., larger than 0.12 bits/pixel), but also maintain a low MER (i.e., less than 0.22) even under voluminous distortions.

REFERENCES

[1] Berghel, H., and L. O'Gorman. "Protecting Ownership Rights through Digital Watermarking." *IEEE Computer Mag.* 29 (1996): 101.

[2] Eggers, J., and B. Girod. *Informed Watermarking.* Norwell, MA: Kluwer Academic, 2002.

[3] Wu, Y. T. "Multimedia Security, Morphological Processing, and Applications." Ph.D. diss., New Jersey Institute of Technology, 2005.

[4] Nikolaidis, N., and I. Pitas. "Robust Image Watermarking in the Spatial Domain." *Signal Processing* 66 (1998): 385.

[5] Celik, M. et al. "Hierarchical Watermarking for Secure Image Authentication with Localization." *IEEE Trans. Image Processing* 11 (2002): 585.

[6] Voyatzis, G., and I. Pitas. "Applications of Toral Automorphisms in Image Watermarking." In *Proc. IEEE Int. Conf. on Image Processing.* Lausanne, Switzerland: IEEE, 1996.

[7] Mukherjee, D., S. Maitra, and S. Acton. "Spatial Domain Digital Watermarking of Multimedia Objects for Buyer Authentication." *IEEE Trans. on Multimedia* 6 (2004): 1.

[8] Wong, P. W. "A Public Key Watermark for Image Verification and Authentication." In *Proc. IEEE Int. Conf. Image Processing.* Chicago, 1998.

[9] Cox, I., et al. "Secure Spread Spectrum Watermarking for Multimedia." *IEEE Trans. Image Processing* 6 (1997): 1673.

[10] Zhao, D., G. Chen, and W. Liu. "A Chaos-Based Robust Wavelet-Domain Watermarking Algorithm." *Chaos, Solitons and Fractals* 22 (2004): 47.

[11] Miller, M. L., G. J. Doerr, and I. J. Cox. "Applying Informed Coding and Embedding to Design a Robust High-Capacity Watermark." *IEEE Trans. Image Processing* 13 (2004): 792.

[12] Wu, Y. T., and Shih, F. Y. "An Adjusted-Purpose Digital Watermarking Technique." *Pattern Recognition* 37 (2004): 2349.

[13] Shih, F. Y., and Y. Wu. "Enhancement of Image Watermark Retrieval Based on Genetic Algorithm." *J. Visual Communication and Image Representation* 16 (2005): 115.

[14] Lin, S. D., and C.-F. Chen. "A Robust DCT-Based Watermarking for Copyright Protection." *IEEE Trans. Consumer Electronics* 46 (2000): 415.

[15] Wu, Y. T., and F. Y. Shih. "Genetic Algorithm Based Methodology for Breaking the Steganalytic Systems." *IEEE Trans. Systems, Man, and Cybernetics,* part B, 36 (2006): 24.

[16] Shih, F. Y., and S. Y. Wu. "Combinational Image Watermarking in the Spatial and Frequency Domains." *Pattern Recognition* 36 (2003): 969.

[17] Wang, H., H. Chen, and D. Ke. "Watermark Hiding Technique Based on Chaotic Map." In *Proc. IEEE Intl. Conf. on Neural Networks and Signal Processing.* Nanjing, China, 2003.

10 Introduction to Digital Steganography

As digital information and data are transmitted over the Internet more often than ever before, new technology for protecting and securing sensitive messages needs to be discovered and developed. Digital steganography is the art and science of hiding information into covert channels, so as to conceal the information and prevent the detection of the hidden message. Information security research on covert channels is generally not a major player, but has been an extremely active topic in academia, industry, and government domains for the past 30 years.

Steganography and cryptography are used in data hiding. Cryptography is the science of protecting data by scrambling so that nobody can read it without given methods or keys; it allows an individual to encrypt data so that the recipient is the only person able to decipher it. Steganography is the science of obscuring the message into a host object (carrier) with the intent of not drawing suspicion to the context in which the message was transferred.

The term *steganography* is derived from two Greek words *steganos*, meaning "covered," and *graphein*, meaning "to write." Its intenion is to hide information in a medium in such a manner that no one except the anticipated recipient knows of its existence. As children we may have experienced writing an invisible message with lemon juice and asking people to find the secret message by holding the paper up against a light bulb. We have thus already, in a sense, used the technology of steganography.

The history of steganography can be traced back to around 440 B.C.E, where the Greek historian Herodotus described in his writings about two events: one used wax to cover secret messages, and the other used shaved heads [1]. In ancient China, military generals and diplomats hid secret messages on thin sheets of silk or paper. One famous story on the successful revolt of the Han Chinese against the Mongolians during the Yuan dynasty tells of a steganographic technique: during the Mid-Autumn Festival, the Han people inserted a message (as hidden information) inside moon cakes (as cover objects) and gave these cakes to their members to deliver information of the planned revolt. Nowadays, with the advent of modern computer technology, a great number of steganographic algorithms take the old-fashioned steganographic techniques and update them in state-of-the-art information hiding, watermarking, and steganography literature.

Modern techniques in steganography have far more powerful tools. Many software tools allow a sender to embed messages in digitized information—typically audio, video, or still image files—that are sent to a recipient. Although steganography has attracted great interest from military and governmental organizations, there is even a great interest shown by commercial companies who wish to safeguard their information from piracy. Today steganography is often used to transmit information safely and embed trademarks securely in images and music files to ensure copyright.

10.1   TYPES OF STEGANOGRAPHY

Steganography can be divided into three types: technical, linguistic, and digital. Technical steganography applies scientific methods to conceal secret messages while linguistic steganography uses written natural language. Digital steganography, developed with the advent of computers, employs computer files or digital multimedia data. In this section, we describe each type respectively.

10.1.1   TECHNICAL STEGANOGRAPHY

Technical steganography applies scientific methods to conceal a secret message, such as the use of invisible ink, microdots, and shaved heads. Invisible ink is the simplest method: one writes secret information on a piece of paper, and when it dries the written information disappears. The paper looks the same just as the original blank piece of paper. Organic compounds such as milk, urine, vinegar, or fruit juice can be used. When heat or ultraviolet light is applied the dried message will become dark and visible to the human eye. In 1641, Bishop John Wilkins invented an invisible ink using onion juice, alum, and ammonia salts. Modern invisible inks fluoresce under ultraviolet light and are used in counterfeit detecting.

With the advent of photography, microfilm was created as a medium of recording a large amount of information in a very small space. In World War II, the Germans used microdots to convey secret information. The technique gained fame when J. Edgar Hoover, then head of the U.S. Federal Bureau of Investigation, called it "the enemy's masterpiece of espionage." The secret message was first photographed, and then reduced to the size of a printed period—a microdot—to be pasted onto innocuous host documents such as newspapers or magazines. Because of their size these microdots would receive little or no attention, but through the transmission of large numbers of them one could hide secret messages, drawings, or even images.

Another version of this type of steganography uses printers, which print a great many small yellow dots that are hardly visible under normal white-light illumination. The pattern of small yellow dots constructs secret messages, and when placed under different colored light (e.g., blue light) the messages become visible.

Steganography using shaved heads started when Histaeus, the ruler of Miletus, planned to send a message to his friend Aristagorus urging his revolt against the Persians. He shaved the head of his most trusted slave as the messenger, and then tattooed a message or a symbol on the messenger's head. After the hair grew back, the messenger ran to Aristagorus to deliver the hidden message. When he arrived, his head was shaved again to reveal the secret message.

Later in Herodotus's history, the Spartans received a secret message from Demeratus, a Greek in exile in Persia, that Xerxes of the Persian army was preparing to invade Greece. Fearing discovery, Demeratus securely concealed the secret message by scraping the wax off one of the writing tablets, inscribing the message on the wood, and then recovering the tablet with wax to make it look typical. These tablets were delivered to the Spartans, who upon receiving the message hastened to save Greece from the Persians.

10.1.2 Linguistic Steganography

Linguistic steganography utilizes written natural language to hide information. It can be categorized into *semagrams* and *open codes*.

Semagrams hide secret messages using visual symbols or signs. Modern updates to this method use computers to hide the secret message. For example, we can alter the font size of specific letters or change their locations, raising or lower them, to conceal messages. We can also add extra spaces in specific locations of the text or adjust the spacing of lines. Semagrams can also be encoded in pictures. The most well-known version of this picture encoding is the secret message in *The Adventure of the Dancing Men* by Sir Arthur Conan Doyle.

Semagrams as described above are sometimes not secure enough. Another secret method is used when spies want to set up meetings or pass information to their networks. This includes hiding information in everyday objects such as newspapers or clothing. Sometimes meeting times can be hidden in reading materials. In one case, an advertisement for a car, placed in a city newspaper during a specific week, read, "Toyota Camry, 1996, needs engine work, sale for $1500." If the spy saw this advertisement, he would know that the Japanese contact person wanted to have personal communication immediately.

Open codes hide secret messages in a specifically designed pattern on the document that is not obvious to the average reader. An example of an open code was used by a German spy in World War II as, "Apparently neutral's protest is thoroughly discounted and ignored. Isman hard hit. Blockade issue affects pretext for embargo on by-products, ejecting suets and vegetable oils." If we collect the second letter in each word, we can extract the secret message as, "Pershing sails from NY June 1."

10.1.3 Digital Steganography

Computer technology has made it a lot easier to hide messages and much more difficult to discover them. Digital steganography is the science of hiding secret messages within digital media, such as digital image, audio, or video files. There are many different methods for digital steganography, including least-significant-bit (LSB) substitution, message scattering, masking and filtering, and various image processing functions.

Secret information or images can be hidden in image files because image files are often very large. When a picture is scanned, a sampling process is conducted to quantize the picture into a discrete set of real numbers. These samples are the gray levels at an equally spaced array of pixels. The pixels are quantized to a set of discrete gray level values, which are also taken to be equally spaced. The result of sampling and quantizing produces a digital image. For example, if we use 8-bit quantization (i.e., 256 gray levels) and 500-line squared sampling, this would generate an array of 250,000 8-bit numbers. It means that a digital image will need two million bits.

We can also hide secret messages on a computer hard drive in a secret partition. The hidden partition is invisible although it can be accessed by disk configuration and other tools. Another form is to use network protocol. By using a secret protocol, which includes a sequence number field (in transmission control segments) and the identification field (in Internet protocol packets), we can construct the covert communication channels.

Modern digital steganographic software employs sophisticated algorithms to conceal secret messages. This reflects a particularly significant present-day threat due to the great number of digital steganography tools freely available on the Internet that can be used to hide any digital file inside of another digital file. With such easy access and simple usage, criminals are inclined to conceal their activities in cyberspace. It has been reported that the Al Qaeda terrorists used digital steganography tools to deliver messages. With the devastating attacks on the Pentagon and New York City's World Trade Center (as well as other proposed locations) on September 11, 2001, there were indications that the Al Qaeda terrorists had used steganography to conceal their correspondence during the planning of these assaults. Therefore, digital steganography presents a grand challenge to law enforcement as well as industry because detecting and extracting hidden information is very difficult.

10.2   APPLICATIONS OF STEGANOGRAPHY

10.2.1   COVERT COMMUNICATION

A famous classic steganographic model presented by Simmons is the prisoners' problem [2], although it was originally introduced to describe a cryptography scenario.

The basic applications of steganography relate to secret communications. Modern computer and networking technology allows individual, group, and company to host a web page that may contain secret information meant for certain users. Anyone can access the web page or download information from it; however, the hidden information is invisible and does not draw any attention. The extraction of the secret information would require specific software with the correct key. Adding encryption to the secret information would further enhance its security. The situation is similar to our often hiding important documents or valuable merchandise in a safe, and/or further security hiding the safe in a secret place that is difficult to access.

All digital data files can be used for steganography, but the files containing a superior degree of redundancy are more appropriate. *Redundancy* is defined as the number of bits needed for an object to provide accurate representation. If we remove the redundant bits, the object will look the same. Digital images and audio files mostly contain a great amount of redundant bits. They are therefore often used as cover objects.

10.2.2   ONE-TIME PAD COMMUNICATION

The *one-time pad* was originated in cryptography to provide a random private key that can only be used once for encrypting a message; we then decrypt the message using a one-time matching pad and key. The communication is generated using a string of numbers or characters having the length as the longest message sent. A random-number generator is used to generate the string of values, which are stored on a pad or a device. The pads are then delivered to the sender and the receiver. Usually, a collection of secret keys is delivered and can only be used once in such a manner that one key is for each day in a month, and will expire at the end of that day.

Messages encrypted by a randomly generated key possess the advantage that there is no solution theoretically to uncover the code by analyzing a series of the messages. All encryptions are different in nature and bear no relationship to each other. This kind of encryption is referred to as *100% noise source* encryption; only the sender and receiver are able to delete the noise. We should note that the one-time pad can be only used once for security reasons. If it were designed to be reused, someone might be able to perform comparisons of multiple messages in order to extract a key for deciphering the messages. One-time pad technology was notably adopted in secret communications during World War II and the Cold War. In today's Internet communication, it is also used in public key cryptography.

The one-time pad is the only encryption method that can be mathematically proven to be unbreakable. The technique can be used to hide an image by splitting it into two random layers of dots. When they are superimposed upon each other, the image appears. The encrypted data or images are perceived as a set of random numbers. It is obvious that no prison warden would want to permit inmates to exchange encrypted, random, malicious-looking messages. Therefore, one-time pad encryption, though statistically secure, will be practically impossible to use in this scenario.

10.3 EMBEDDING SECURITY AND IMPERCEPTIBILITY

A robust steganographic system must be extremely secure against all kinds of attacks (i.e., steganalysis), and must not degrade the visual quality of cover images. We can combine several strategies to ensure security and imperceptibility. For security we can use the chaotic mechanism, frequency hopping structure, pseudorandom number generator, or patchwork locating algorithm. For imperceptibility we can use set partitioning in hierarchical trees coding, discrete wavelet transform, and parity check embedding.

Benchmarking of steganographic techniques is a complicated task that requires examination of a set of mutually dependent performance indices [3–9]. A benchmarking tool should be able to interact with different performance aspects such as visual quality and robustness. The input images of the benchmarking system should be ones that vary in size and frequency since these factors affect system performance.

We should also evaluate the execution time of the embedding and detection modules. Mean, maximum, and minimum execution times for the two modules might be evaluated over the set of all keys and messages for each host image. We should measure the perceptual quality of stego-images by subjective quality evaluation and by a quantitative way that correlates well with the way human observers perceive image quality. We should also evaluate the maximum number of information bits that can be embedded per host image pixel.

Since steganography is expected to be robust against host-image manipulations, tests for judging the performance of a steganographic technique when applied to distorted images constitute an important part of a benchmarking system. In order to make embedded messages undetectable, the set of attacks available should include all operations that the average user or an intelligent pirate might employ. It should also include signal processing operations and distortions that occur during normal image usage, transmission, storage, and the like.

10.4   EXAMPLES OF STEGANOGRAPHIC SOFTWARE

Steganography embeds secret messages into innocuous files. Its goal is not only to hide the message but to also let the stego-image pass without causing suspicion. There are many steganographic tools available for hiding messages in image, audio, and video files. In this section, we briefly introduce S-Tools, StegoDos, EzStego, and JSteg-Jpeg.

10.4.1   S-Tools

The S-Tools program [10], standing for Steganography Tools, was written by Andy Brown to hide the secret messages in BMP, GIF, and WAV files. It is a combined steganographic and cryptographic product, since the messages to be hidden are encrypted using symmetric keys. It uses LSB substitution in files that employ loss-less compression, such as 8- or 24-bit color- and pulse-code modulation, and applies a pseudorandom number generator to make the extraction of secret messages more difficult. It also provides for the encryption and decryption of hidden files with several different encryption algorithms.

We can open up a copy of S-Tools and drag images and audios into it. To hide files we can drag them over open audio or image windows. We can hide multiple files in one audio or image, and compress the data before encryption. Multithreaded operation provides the flexibility of many hide/reveal operations running simultaneously without interfering with other work. We can also close the original image or audio file without affecting the ongoing threads.

10.4.2   StegoDos

StegoDos [11], also known as Black Wolf's Picture Encoder, consists of a set of programs that allow us to capture an image, encode a secret message, and display the stego-image. This stego-image may also be captured again into another format using a third-party program. Then we can recapture it and decode the hidden message. It only works with 320×200 images with 256 colors. It also uses LSB substitution to hide messages.

10.4.3   EzStego

EzStego, developed by Romana Machado, simulates invisible ink in Internet communications. It hides an encrypted message in a GIF format image file by modulating LSBs. It begins by sorting the colors in the palette so that the closest colors fall next to each other. Then similar colors are paired up, and for each pair one color will represent 1 while the other will represent 0. It encodes the encrypted message by replacing the LSB. EzStego compares the bits it wants to hide to the color of each pixel. If the pixel already represents the correct bit, it remains untouched; otherwise, its color will be changed to its pair color. The changes in LSBs are so small that they are undetectable to the human eye. EzStego treats the colors as sites in the three-dimensional red-green-blue space, and intends to search for the shortest path through all of the sites.

10.4.4 JSTEG-JPEG

Jsteg-Jpeg, developed by Derek Upham, can read multiple format images and embed the secret messages to be saved as JPEG images [12]. It utilizes the splitting of JPEG encoding into lossy and lossless stages. The lossy stages use DCT and a quantization step to compress the image data, and the lossless stage uses Huffmann coding to compress the image data. DCT is a compression technique since the cosine values cannot be accurately calculated and the repeated calculations can introduce rounding errors. It therefore inserts the secret message into the image data between these two steps. It also modulates the rounding processes in the quantized DCT coefficients. The image degradation is affected by the embedded amount as well as the quality factor used in JPEG compression. The trade-off between the two can be adjusted to allow the imperceptible requirement.

REFERENCES

[1] Wrixon, F. B. *Codes, Ciphers and Other Cryptic and Clandestine Communication.* New York: Black Dog and Leventhal, 1998.

[2] Simmons, G. J. "The Prisoners' Problem and the Subliminal Channel." In *Proc. Intl. Conf. Advances in Cryptology.* 1983.

[3] Fei, C., D. Kundur, and R. H. Kwong. "Analysis and Design of Secure Watermark-Based Authentication Systems." *IEEE Trans. Information Forensics and Security* 1 (2006): 43.

[4] Lyu, S., and H. Farid. "Steganalysis Using Higher-Order Image Statistics." *IEEE Trans. Information Forensics and Security* 1 (2006): 111.

[5] Sullivan, K., et al. "Steganalysis for Markov Cover Data with Applications to Images." *IEEE Trans. Information Forensics and Security* 1 (2006): 275.

[6] Johnson, N., and S. Jajodia, S. "Exploring Steganography: Seeing the Unseen." *IEEE Computer* 31, (1998): 26.

[7] Johnson, N. F., Z. Durric, and S. Jajodia. *Information Hiding: Steganography and Watermarking*, Boston: Kluwer Academic, 2001.

[8] Kharrazi, M., H. T. Sencar, and N. D. Memon. "Benchmarking Steganographic and Steganalysis Techniques." In *Proc. Intl. Conf. Security, Steganography, and Watermarking of Multimedia Contents.* San Jose, CA, 2005.

[9] Wu, Y. T., and F. Y. Shih. "Genetic Algorithm Based Methodology for Breaking the Steganalytic Systems." *IEEE Trans. Systems, Man, and Cybernetics,* part B, 36 (2006): 24.

[10] Brown, A. *S-Tools for Windows.* Shareware application, 1994. Available online at ftp://idea.sec.dsi.unimi.it/pub/security/crypt/code/s-tools4.zip.

[11] Wolf, B. *StegoDos—Black Wolf's Picture Encoder* v0.90B, public domain. Available online at ftp://ftp.csua.berkeley.edu/pub/cypherpunks/steganography/stegodos.zip.

[12] Korejwa, J. Jsteg Shell 2.0. Application. Available online at http://www.tiac.net/users/korejwa/steg.htm.

11 Steganalysis
Attacks against Steganography

The goal of steganography is to avoid drawing suspicion to the transmission of a secret message. Its essential principle requires that the stego-image (i.e., the object containing hidden messages) should be perceptually indistinguishable to the degree that suspicion is not raised. The process of detecting the secret message, called *steganalysis*, relies on the fact that hiding information in digital media alters the characteristics of the carriers and places some sort of distortion to the carriers. By analyzing various image features between stego-images and cover images (i.e., the images containing no hidden messages), a steganalytic system is developed to detect stego-images. This chapter examines steganalysis. Section 11.1 gives an overview, and section 11.2 examines the statistical properties of images. The visual steganalytic system is discussed in section 11.3, while section 1.4 examines a steganalytic system based on image quality measure (IQM). Section 11.5 looks at learning strategies, and section 11.6 examines a steganalytic system within the frequency domain.

11.1 AN OVERVIEW

Digital steganalytic systems can be categorized into two classes: the *spatial-domain steganalytic system* (SDSS) and the *frequency-domain steganalytic system* (FDSS). The SDSS is adopted for checking lossless compressed images by analyzing the statistical features of the spatial domain [1, 2]. The simplest SDSS is the least-significant-bit (LSB) substitution. This technique utilizes the concept that the LSB is considered to be random noise, so its modification will not affect the overall image visualization. For lossy compression images (such as JPEG images), the FDSS is used to analyze the statistical features of the frequency domain [3, 4]. Westfeld and Pfitzmann have presented two SDSSs based on visual and chi-square attacks [1]. The visual attack uses human eyes to inspect stego-images by checking their lower bit planes. The chi-square attack can automatically detect the specific characteristic generated by the LSB steganographic technique. Avcibas, Memon, and Sankur have proposed the steganalysis using *image quality measure* (IQM), which is based on a hypothesis that the steganographic systems leave statistical evidences that can be exploited for detection using IQM and multivariate regression analysis [2]. Fridrich, Goljan, and Hogea have presented an FDSS for detecting JPEG stego-images by analyzing their discrete cosine transformation (DCT) with cropped images [3].

Steganalysis is the process of detecting steganography. Basically, there are two methods of ascertaining the presence of modified files. One is called *visual analysis*, which involves comparing a suspected modified file with the original file. It intends to reveal the presence of secret communication through inspection, either by eyes or with the help of a computer system, typically decomposing the image into its bit planes. Although this method is very simple, it is not very effective as, most of the time, the original copy is unavailable.

The other is called *statistical analysis*, which detects changes in patterns of the pixels and the frequency distribution of the intensities. This analysis can detect whether an image was modified by checking to see if its statistical properties deviate from a norm. Therefore, it intends to find out even slight alterations in the statistical behavior caused by steganographic embedding.

Steganalysis techniques must be updated frequently and evolve continuously in order to combat new and more sophisticated steganographic methods. Within U.S. law enforcements agencies, for example, there is no standard steganographic guideline available, but there are certain rules that investigators adopt when they are looking for hints that may suggest the use of steganography. These include the technical capabilities of the computer's owner, software clues, examination of program files and multimedia files, and types of crimes.

11.2   THE STATISTICAL PROPERTIES OF IMAGES

There are statistical properties in natural scene images for human perception. Wavelengths of light generate visual stimuli. When a wavelength dynamically changes, the perceived color varies from red, orange, yellow, green, and blue to violet. Human visual sensitivities are closely related to color; for example, we can see light in the turquoise-green-yellow portion of the color spectrum more easily than red or blue. We can divide the composition of a color into three components: brightness, hue, and saturation. The eye's adaptation to a particular hue distorts the perception of other hues; for example, gray seen after green, or seen against a green background, looks purplish.

Researchers in statistical steganalysis have put forth efforts toward advancing their methods due to the increasing preservation of first-order statistics in steganography to avoid detection. Encryption of the hidden message also makes detection more difficult because the encrypted data generally have a high degree of randomness [4, 5]. In addition to detection, the recovery of the hidden message appears to be a complicated problem since it requires knowledge of the cryptoalgorithm or the encryption key [6].

LSB insertion in a palette-based image will produce a great number of duplicate colors. Some identical (or nearly identical) colors may appear twice in the palette. Some steganographic techniques that rearrange the color palette in order to hide messages may cause structural changes, which will generate a signature of the steganography algorithm [7, 8].

Similar colors tend to be clustered together. Changing the LSB of a color pixel (e.g., in LSB steganography) does not degrade overall visual perception. There are four other steganalytic methods, including the Reed–Solomon (RS) method; pairs of values method; pairs analysis method; and sample pair method, which take advantage of statistical changes by modeling changes in certain values as a function of the percentage of pixels within embedded steganography. We briefly describe each method below.

The RS method has been developed by Fridrich, Goljan, and Du [9], who divide a cover image into a set of 2×2 disjoint groups instead of 4×1 slices. A discrimination function is applied to quantify the smoothness or regularity of the group of pixels. The variation is defined as the sum of the absolute value of the difference between adjacent pixels in the group. An invertible operation, called *shifted LSB flipping*, is defined as a permutation that consists entirely of two cycles.

The pairs of values method has been developed by Pfitzman and Westfeld [10], who use a fixed set of pairs of values (PoVs) to flip for embedding message bits. They developed an algorithm called FlipEmbed, which simply flips the LSB of each pixel in the cover image to match the message bit. They construct PoVs using quantized DCT coefficients, pixel values, or palette indexes. We spread out two values from each pair irregularly in the cover image before message embedding. The occurrences of the values in each pair are inclined to be equal after embedding. The sum of occurrences of a pair of colors in the image remains the same if we swap one value for another. This concept is used in designing a statistical chi-square test.

The pairs analysis method has been developed by Fridrich, Goljan, and Soukal [11]. It can detect steganography with diluted changes, uniformly straddled over the whole carrier medium. It can also estimate the embedded message length. The pairs of colors, which are exchanged during LSB embedding, correspond to a preordering of the palette via LSB steganography.

The sample pair method has been developed by Dumitrescu, Wu, and Wang [12]. In this method the image is represented as a set of sample pairs, which are divided into submultisets according to the difference between the pixel value before and after LSB embedding. This assumes that the probability of sample pairs having a large even component is equal to that having a large odd component for natural images. Therefore, the accuracy of the method depends upon the accuracy of this assumption. Dumitrescu et al. use finite-state machines to statistically measure the cardinalities of the trace multisets before and after LSB embedding.

Steganographic techniques generally modify the statistical properties of the carrier; a longer hidden message will alter its carrier more than a shorter one will [4, 13–15]. Statistical analysis is utilized to detect hidden messages, particularly in blind steganography [8]. Statistical analysis of images can determine whether their statistical properties deviate from the expected norm [4, 5, 15]. These include means, variances, and chi-square (χ^2) tests [7].

11.3 THE VISUAL STEGANALYTIC SYSTEM

Kessler has presented an overview of steganography for the computer forensics examiner [16]. The steganalysis strategy usually follows the method that the steganography algorithm uses. A simple way is to visually inspect the carrier and steganography media. Many simple steganographic techniques embed secret messages in the spatial domain and select message bits in the carrier that are independent of its content [7].

Most steganographic systems embed message bits in a sequential or pseudorandom manner. If the image contains certain connected areas of uniform color or saturated color (i.e., 0 or 255), we can use visual inspection to find suspicious artifacts. When the artifacts are not found, an inspection of the LSB bit plane is conducted.

Westfeld and Pfitzmann have presented a visual steganalytic system that uses an assignment function of color replacement called the *visual filter* [10]. The idea is to remove all parts of the image that contain a potential message. The filtering process relies on the presumed steganographic tool and can produce the attacked carrier medium steganogram, the extraction of potential message bits, and the visual illustration.

11.4   IQM-BASED STEGANALYTIC SYSTEM

Image quality can be characterized by correlation between the display and the human visual system (HVS). With the advent of IQM, we aim at minimizing subjective decisions. Computational image quality analysis is geared toward the efforts of developing an accurate computational model for describing image quality. There are two types of IQM: full-referenced IQM and no-reference IQM. Full-referenced IQM has the original (i.e., undistorted) image available as the reference to be compared against the distorted image. It applies different kinds of distance metrics to measure the closeness between them. The no-reference IQM takes the distorted image alone for analysis. It uses image processing operators, such as an edge detector, to compute the quality index.

Avcibas et al. have developed a technique based on IQM for steganalysis of the images that have been potentially subjected to steganographic algorithms, both within the passive-warden and active-warden frameworks [2]. Note that the passive warden is to inspect the message in order to determine whether it contains a hidden message and then to conduct a proper action. On the other hand, the active warden is always to alter the messages no matter what trace of a hidden message may or may not appear. They assume that steganographic schemes leave statistical evidence that can be exploited for detection with the aid of image quality features and multivariate regression analysis. To this effect, image quality metrics have been identified based on the analysis-of-variance technique as feature sets to distinguish between cover images and stego-images. The classifier between cover and stego-images is built using multivariate regression on the selected quality metrics and is trained based on an estimate of the original image.

We can generally categorize IQMs into three types, as follows. Let $F(j,k)$ denote the pixel value in row j and column k of a reference image of size $M \times N$, and $\widehat{F}(j,k)$ denote a testing image.

Type i: IQMs Based on Pixel Distance

1. Average Distance

$$AD = \sum_{j=1}^{M} \sum_{k=1}^{N} (F(j,k) - \widehat{F}(j,k))/MN \qquad (11.1)$$

2. L2 (Euclidean) Distance

$$L2D = \frac{1}{MN}\left(\sum_{j=1}^{M}\sum_{k=1}^{N}(F(j,k)-\widehat{F}(j,k))^2\right)^{1/2} \tag{11.2}$$

Type 2: IQMs Based on Correlation

3. Structure Content

$$SC = \sum_{j=1}^{M}\sum_{k=1}^{N}F(j,k)^2 \bigg/ \sum_{j=1}^{M}\sum_{k=1}^{N}\widehat{F}(j,k)^2 \tag{11.3}$$

4. Image Fidelity

$$IF = 1-\left(\sum_{j=1}^{M}\sum_{k=1}^{N}(F(j,k)-F(j,k)\right)^2 \bigg/ \sum_{j=1}^{M}\sum_{k=1}^{N}F(j,k)^2 \tag{11.4}$$

5. N (Normalized) Cross-Correlation

$$NK = \sum_{j=1}^{M}\sum_{k=1}^{N}F(j,k)\widehat{F}(j,k) \bigg/ \sum_{j=1}^{M}\sum_{k=1}^{N}F(j,k)^2 \tag{11.5}$$

Type 3: IQMs Based on Mean Square Error

6. Normal Mean Square Error

$$NMSE = \sum_{j=1}^{M}\sum_{k=1}^{N}(F(j,k)-\widehat{F}(j,k))^2 \bigg/ \sum_{j=1}^{M}\sum_{k=1}^{N}F(j,k)^2 \tag{11.6}$$

7. Least Mean Square Error (LMSE)

$$LMSE = \sum_{j=1}^{M-1}\sum_{k=2}^{N-1}(F(j,k)-\widehat{F}(j,k))^2 \bigg/ \sum_{j=1}^{M-1}\sum_{k=2}^{N-1}O(F(j,k))^2, \tag{11.7}$$

where $O(F(j,k)) = F(j+1,k)+F(j-1,k)+F(j,k+1)+F(j,k-1)-4F(j,k)$

8. PMSE (Peak Mean Square Error)

$$PMSE = \frac{1}{MN}\sum_{j=1}^{M}\sum_{k=1}^{N}[F(j,k)-\hat{F}(j,k)]^2 \bigg/ \{\max_{j,k}[F(j,k)]\}^2 \tag{11.8}$$

9. PSNR (Peak Signal to Noise Ratio)

$$PSNR = 20 \times \log_{10} \left\{ 255 \Big/ \left\{ \sum_{j=1}^{M} \sum_{k=1}^{N} [F(j,k) - \hat{F}(j,k)]^2 \right\}^{1/2} \right\}$$ (11.9)

Let $f(x_i)$ denote the IQM score of an image under the degree of distortion x_i. The IQMs used for measuring image blurs must satisfy the monotonically increasing or decreasing property. That is, if $x_{i+1} > x_i$, then $f(x_{i+1}) - f(x_i) > 0$ or $f(x_{i+1}) - f(x_i) < 0$. The sensitivity of IQM is defined as the score of the aggregate relative distance as:

$$\sum_{i=1}^{7} \frac{f(x_{i+1}) - f(x_i)}{f(x_i)}$$ (11.10)

A quality measure called *mean opinion score* has been proposed by Grgic, Grgic, and Mrak to correlate with various IQMs. However, it requires human interaction and is not suited for automatic processes [17]. To measure the degree of "closeness" to HVS, we develop the following procedure for measuring HVS as below, where Nill's band-pass filter is utilized [18]:

1. Perform DCT on the original image $C(x,y)$ and the distorted image $\hat{C}(x,y)$ to obtain $D(u,v)$ and $\hat{D}(u,v)$, respectively.
2. Convolve $D(u,v)$ and $\hat{D}(u,v)$ with the following band-pass filter to obtain $E(u,v)$ and $\hat{E}(u,v)$:

$$H(p) = \begin{cases} 0.05 e^{p^{0.554}}, & \text{if } p < 7 \\ e^{-9|\log_{10} p - \log_{10} 9|^{2.3}}, & \text{if } p \geq 7 \end{cases}, \text{ where } p = (u^2 + v^2)^{1/2}.$$ (11.11)

3. Perform inverse DCT of $E(u,v)$ and $\hat{E}(u,v)$ to obtain $F(x,y)$ and $\hat{F}(x,y)$.
4. Perform Euclidean distance computation on $F(x,y)$ and $\hat{F}(x,y)$ as

$$d = \left\{ \sum_{x} \sum_{y} [F(x,y) - \hat{F}(x,y)]^2 \right\}^{1/2}$$ (11.12)

11.5 LEARNING STRATEGIES

There are many types of learning strategies, including statistical learning; neural networks; and the support vector machine (SVM), which is used to generate an optimal separating hyperplane by minimizing the generalization error without using the assumption of class probabilities such as a Bayesian classifier [19–22]. The decision hyperplane of SVM is determined by the most informative data instances, called support vectors (SVs). In practice, these SVs are a subset of the entire training data. By applying feature reduction in both input and feature space, a fast nonlinear SVM can be designed without a noticeable loss in performance [23, 24].

The training procedure of SVMs usually requires huge memory space and significant computation time due to the enormous amounts of training data and the quadratic programming problem. Some researchers proposed incremental training or active learning to shorten the training time [25–30]. The goal is to select a subset of training samples while preserving the performance as using all the training samples. Instead of learning from randomly selected samples, the active learning queries the informative samples in each incremental step.

Syed, Liu, and Sung have developed an incremental learning procedure by partitioning the training dataset into subsets [26]. At each incremental step, only support vectors are added to the training set for the next step. Campbell, Cristianini, and Smola have proposed a query learning strategy by iteratively requesting the label of a sample that is closest to the current hyperplane [27]. Schohn and Cohn have presented an optimal greedy strategy by calculating class probability and expected error [25]. A simple heuristic is used for selecting training samples according to their proximity with respect to the dividing hyperplane.

Moreover, An, Wang, and Ma have developed an incremental learning algorithm by extracting initial training samples as the margin vectors of two classes, selecting new samples according to their distances to the current hyperplane, and discarding the training samples that do not contribute to the separating plane [28]. Nguyen and Smeulders have pointed out that prior-data distribution is useful for active learning, and that new training data selected from the samples have the maximal contribution to the current expected error [29].

Most existing methods of active learning perform the measurement of proximity to the separating hyperplane. Mitra, Murthy, and Pal have presented probabilistic active learning using a distribution determined by the current separating hyperplane and confidence factor [30]. Unfortunately, they do not provide criteria for choosing the test samples. There are two issues in their algorithm. First, the initial training samples are selected randomly, which may generate the initial hyperplane far away from the optimal solution. Second, using only the support vectors from the previous training set may lead to bad performance.

11.5.1   INTRODUCTION OF THE SUPPORT VECTOR MACHINE

Cortes and Vapnik have introduced the two-group SVM into classification [31]. The SVM was originally designed as a binary classifier. Researchers have proposed extensions to multiclass (or multigroup) classification by combining multiple SVMs, such as one-against-all, one-against-one, and the directed acyclic graph SVM (DAG-SVM) [32]. The one-against-all method involves learning k binary SVMs during the training stage. For the i-th SVM, we label all training samples by $y(j)_{j=i} = +1$ and $y(j)_{j \neq i} = -1$. In the testing, the decision function is obtained by

$$\text{Class of } \bar{\mathbf{x}} = \arg\max_{i=1,2,\dots,k} ((w^i)^T \phi(\bar{\mathbf{x}}) + b^i), \qquad (11.13)$$

where $(w^i)^T \phi(\bar{\mathbf{x}}) + b^i = 0$ represents the hyperplane for the i-th SVM. It has been shown that we can solve the k-group problem simultaneously [33–34].

The one-against-one method constructs $k(k-1)/2$ SVMs in a tree structure. Each SVM represents a distinguishable pairwise classifier from different classes in the training stage. In the testing, each leaf SVM pops up one desirable class label, which propagates to the upper level in the tree until it processes the root SVM of the tree.

The DAG-SVM combines multiple binary one-against-one SVM classifiers into one multiclass classifier [32]. It has $k(k-1)/2$ internal SVM nodes, but in the testing phase it starts from the root node and moves to the left or right subtree depending on the output value of the binary decision function until a leaf node is reached that indicates the recognized class. There are two main differences between the one-against-one and DAG methods. One is that one-against-one needs to evaluate all $k(k-1)/2$ nodes, while DAG only needs to evaluate k nodes due to the different testing phase schema. The other is that one-against-one uses a bottom-up approach, but the DAG uses a top-down approach.

Suppose we have l patterns, and each pattern consists of a pair: a vector $\mathbf{x}_i \in \mathbf{R}^n$ and the associated label $y_i \in \{-1, 1\}$. Let $X \subset R^n$ be the space of patterns, $Y = \{-1, 1\}$ be the space of labels, and $p(\mathbf{x}, y)$ be a probability density function on $X \times Y$. The machine learning problem consists of finding a mapping: $sign(f(\mathbf{x}_i)) \rightarrow y_i$, which minimizes the error of misclassification. Let f denote the decision function. The $sign$ function is defined as

$$sign(f(\mathbf{x})) = \begin{cases} 1 & f(\mathbf{x}) \geq 0 \\ -1 & \text{otherwise.} \end{cases} \tag{11.14}$$

The expected value of the actual error is given by

$$E[e] = \int \frac{1}{2} |y - sign(f(\mathbf{x}))| p(\mathbf{x}, y) d\mathbf{x}. \tag{11.15}$$

However, $p(\mathbf{x}, y)$ is unknown in practice. We use the training data to derive the function f whose performance is controlled by two factors: the training error and the capacity measured by the Vapnik–Chervonenkis (VC) dimension [22]. Let $0 \leq \eta \leq 1$. The following bound holds with a probability of $1 - \eta$:

$$e \leq e_t + \sqrt{\left(\frac{h(\log(2l/h) + 1) - \log(\eta/4)}{l} \right)}, \tag{11.16}$$

where h is the VC dimension and e_t is the error tested on the training set

$$e_t = \frac{1}{2l} \sum_{i=1}^{l} |y_i - sign(f(\mathbf{x}_i))|. \tag{11.17}$$

For binary classification, SVMs use a hyperplane that maximizes the margin (i.e., the distance between the hyperplane and the nearest sample of each class). This hyperplane is viewed as the optimal separating hyperplane (OSH). The training patterns are linearly separable if there exists a vector $\mathbf{w}$ and a scalar b such that the inequality

$$y_i(\mathbf{w}.\mathbf{x}_i + b) \geq 1, \; i = 1, 2, \ldots, l. \tag{11.18}$$

is valid. The margin between the two classes can be simply calculated as

$$m = \frac{2}{|\mathbf{w}|}. \tag{11.19}$$

Therefore, the OSH can be obtained by minimizing $|\mathbf{w}|^2$ subject to the constraint in equation 11.18.

The positive Lagrange multipliers $\alpha_i \geq 0$ are introduced to form the Lagrangian as

$$L = \frac{1}{2}|\mathbf{w}|^2 - \sum_{i=1}^{l}\alpha_i y_i(\mathbf{x}_i \cdot \mathbf{w} + b) + \sum_{i=1}^{l}\alpha_i. \tag{11.20}$$

The optimal solution is determined by the saddle point of this Lagrangian, where the minimum is taken with respect to $\mathbf{w}$ and b, and the maximum is taken with respect to the Lagrange multipliers α_i. By taking the derivatives of L with respect to $\mathbf{w}$ and b, we obtain

$$\mathbf{w} = \sum_{i=1}^{l}\alpha_i y_i \mathbf{x}_i, \tag{11.21}$$

$$\sum_{i=1}^{l}\alpha_i y_i = 0. \tag{11.22}$$

Substituting equations 11.21 and 11.22 into equation 11.20 we obtain

$$L = \sum_{i=1}^{l}\alpha_i - \frac{1}{2}\sum_{i=1}^{l}\sum_{j=1}^{l}\alpha_i\alpha_j y_i y_j \mathbf{x}_i \cdot \mathbf{x}_j, \tag{11.23}$$

Thus, constructing the optimal hyperplane for the linear SVM is a quadratic programming problem that maximizes equation 11.23 subject to equation 11.22.

If the data are not linearly separable in the input space, a nonlinear transformation function $\Phi(\cdot)$ is used to project $\mathbf{x}_i$ from the input space $\mathbf{R}^n$ to a higher dimensional feature space $\mathbf{F}$. The OSH is constructed in the feature space by maximizing the margin between the closest points $\Phi(\mathbf{x}_i)$. The inner-product between two projections is defined by a kernel function $K(\mathbf{x}, \mathbf{y}) = \Phi(\mathbf{x}) \cdot \Phi(\mathbf{y})$. The commonly used kernels include polynomial kernel, Gaussian radial basis function kernel, and sigmoid kernel. Similar to linear SVM, constructing the OSH for SVM with kernel $K(\mathbf{x}, \mathbf{y})$ involves maximizing

$$L = \sum_{i=1}^{l} \alpha_i - \frac{1}{2} \sum_{i=1}^{l} \sum_{j=1}^{l} \alpha_i \alpha_j y_i y_j K(\mathbf{x}_i, \mathbf{x}_j), \tag{11.24}$$

subject to equation 11.22.

In practical applications, the training data in two classes are not often completely separable. In this case, a hyperplane that maximizes the margin while minimizing the number of errors is desirable. It is obtained by maximizing equation 11.24 subject to the constraints $\sum_{i=1}^{l} \alpha_i y_i = 0$ and $0 \leq \alpha_i \leq C$, where α_i is the Lagrange multiplier for pattern $\mathbf{x}_i$ and C is the regularization constant that manages the tradeoff between the minimization of the number of errors and the necessity to control the capacity of the classifier.

11.5.2 Neural Networks

A layered neural network is a network of neurons organized in the form of layers. The first layer is the input layer; in the middle there are one or more hidden layers; and the last layer is the output layer. The architecture of a layered neural network is shown in Figure 11.1. The function of the hidden neurons is to intervene between the external input and the network output. The output vector $\vec{\mathbf{Y}}(\vec{\mathbf{x}})$ is

$$\vec{\mathbf{Y}}(\vec{\mathbf{x}}) = \sum_{i=1}^{l} f_i(\vec{\mathbf{x}}) w_{ik} = \vec{\mathbf{F}} \cdot W \tag{11.25}$$

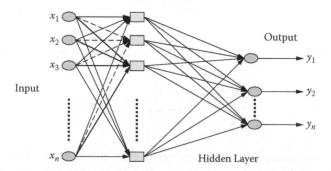

FIGURE 11.1 The architecture of a layered neural network.

where $\vec{F}$ is the hidden layer's output vector corresponding to a sample $\vec{x}$, W is the weight matrix with dimensions of $l \times k$, l is the number of hidden neurons, and k is the number of output nodes. In such a structure, all neurons cooperate together to generate a single output vector. In other words, the output does not only depend on each individual neuron, but on all of them.

There are two categories in neural networks. One is *supervised learning* and the other is *unsupervised learning*. For supervised learning, the target responses are known for the training set, but for unsupervised learning, we do not know the desired responses. This chapter uses supervised learning.

11.5.3 PRINCIPLE COMPONENT ANALYSIS

Principle component analysis (*PCA*) is the suitable data representation in a least-square sense for classical recognition techniques [35]. It can reduce the dimensionality of original images dramatically and retain most information. Suppose we have a population of n faces as the training set $\vec{z} = (\vec{z}_1, \vec{z}_2, ..., \vec{z}_n)^T$, where each component $\vec{z}_i$ represents a training image vector. The mean vector of this population is computed by

$$\bar{\mu} = \frac{1}{N} \sum_{i=1}^{N} \vec{z}_i \tag{11.26}$$

and the covariance matrix is computed by

$$S = \frac{1}{N-1} \sum_{i=1}^{N} (\vec{z}_i - \bar{\mu})(\vec{z}_i - \bar{\mu})^T \tag{11.27}$$

Using the covariance matrix, we calculate eigen vectors and eigen values. Based on nonzero eigen values in a decreasing order, we use the corresponding eigen vectors to construct the eigen space U (also known as the feature space) as $U = (\vec{e}_1, \vec{e}_2, ..., \vec{e}_k)$, where $\vec{e}_j$ is the *j-th* eigen vector. Then we project all images onto this eigen space to find new representation for each image. The new representation is computed by

$$\vec{x}_i = U^T (\vec{z}_i - \bar{\mu}) \tag{11.28}$$

Through the PCA process we can reduce original images from a high-dimensional input space to a much lower dimensional feature space. Note that the PCA transformation can preserve the structure of original distribution without degrading classification.

11.6 THE FREQUENCY-DOMAIN STEGANALYTIC SYSTEM

Like watermarking, steganography can embed hidden messages into the frequency domain through the use of its redundancies. The simplest method is LSB substitution. There are various methods of changing the frequency coefficients for reducing the statistical artifacts in the embedding process—for example, the F5 and Outguess algorithms [36–37].

Westfeld has presented the F5 algorithm to achieve robust and high-capacity JPEG steganography [36]. He embeds messages by modifying the DCT coefficients. The F5 algorithm has two important features: it permutes the DCT coefficients before embedding, and it employs matrix embedding. The first operation has the effect of spreading the changed coefficients evenly across the entire image. The second operation uses non-zero, non-DC coefficients and message length to calculate suitable matrix embedding, which can minimize the number of changes to the cover image. Although the algorithm alters the histogram of DCT coefficients, Westfeld has verified that some characteristics, such as the monotonicity of increments and histograms, are in fact preserved.

Fridrich, Goljan, and Hogea have presented a steganalytic method to detect hidden messages by using the F5 algorithm in JPEG images [38], which estimates the histogram of the cover image from the stego-image. This can be accomplished via stego-image decompression, cropping it by four pixels in both directions to remove the quantization in the frequency domain and recompressing it using the same quality factor as the stego-image.

OutGuess is a steganographic system available in UNIX. The first version (0.13b) is vulnerable to statistical analysis, while the second version (0.2) incorporates the preservation of statistical properties [38]. OutGuess embeds the message into the selected DCT coefficients using a pseudorandom number generator. This generator and a stream cipher are created via a user-defined pass phrase. Basically, there are two steps in Outguess to embed the message. The system first detects the redundant DCT coefficients that have the least effect on the cover image. It then selects bits in which to embed the message according to the obtained information.

Recent intelligence indicates that terrorists may communicate secret information with each other using steganography, hiding messages within images on the Internet. In order to analyze steganographic messaging on the Internet, Provos and Honeyman have presented several tools to retrieve and perform automatic analysis on these images [5]. Alturki and Mersereau have proposed quantizing the coefficients in the frequency domain for embedding secret messages [39]. They first decorrelate the image by randomly scrambling the pixels, which in effect whitens the frequency domain of the image and increases the number of transform coefficients in the frequency domain, thus increasing the embedding capacity. The frequency coefficients are then quantized to even or odd multiples of the quantization step size in order to embed zeros or ones. An inverse discrete Fourier transform is then taken and descrambled. The resulting image is visually incomparable to the original image.

REFERENCES

[1] Westfeld, A., and A. Pfitzmann. "Attacks on Steganographic Systems Breaking the Steganographic Utilities EzStego, Jsteg, Steganos, and S-Tools and Some Lessons Learned." In *Proc. Int. Workshop Information Hiding*. Dresden, Germany, 1999, 61.

[2] Avcibas, I., N. Memon, and B. Sankur. "Steganalysis Using Image Quality Metrics." *IEEE Trans. Image Processing* 12 (2003): 221.

[3] Fridrich, J., M. Goljan, and D. Hogea. "New Methodology for Breaking Steganographic Techniques for JPEGs." In *Proc. SPIE*, Santa Clara, CA: SPIE, 2003.

[4] Farid, H. *Detecting Steganographic Message in Digital Images.* Technical Report TR2001-412, Computer Science. Hanover, NH: Dartmouth College, 2001.

[5] Provos, N. and P. Honeyman. *Detecting Steganographic Content on the Internet.* Technical Report 01-11. Ann Arbor, MI: Center for Information Technology Integration, University of Michigan, 2001.

[6] Fridrich, J., et al. "Quantitative Steganalysis of Digital Images: Estimating the Secret Message Length." *Multimedia Systems* 9 (2003): 288.

[7] Wayner, P. *Disappearing Cryptography: Information Hiding: Steganography and Watermarking.* 2d ed. San Francisco: Morgan Kaufmann, 2002.

[8] Jackson, J. T., et al. "Blind Steganography Detection Using a Computational Immune System: A Work in Progress." *Int. J. Digital Evidence* (2003).

[9] Fridrich, J., M. Goljan, and R. Du. "Reliable Detection of LSB Steganography in Color and Grayscale Images." In *Proc. ACM Workshop Multimedia Security*, Ottawa, Ontario, 2001.

[10] Westfeld, A., and A. Pfitzmann. *Attacks on Steganographic Systems.* Lecture Notes in Computer Science 1768. Berlin: Springer-Verlag, 2000.

[11] Fridrich, J., M. Goljan, and D. Soukal. "Higher-Order Statistical Steganalysis of Palette Images." In *Proc. SPIE Conf. Security and Watermarking of Multimedia Contents* 5. 2003.

[12] Dumitrescu, S., X. Wu, and Z. Wang. "Detection of LSB Steganography via Sample Pair Analysis." In *Proc. Information Hiding Workshop.* 2002.

[13] Fridrich, J., and R. Du. "Secure Steganographic Methods for Palette Images." In *Proc. Information Hiding Workshop.* Dresden, Germany, 1999.

[14] Fridrich, J., and M. Goljan, M. "Practical Steganalysis of Digital Images: State of the Art." In *Proc. SPIE Security and Watermarking of Multimedia Contents* 5. San Jose, CA: SPIE, 2002.

[15] Ozer, H., et al. "Steganalysis of Audio Based on Audio Quality Metrics." In *Proc. SPIE, Security and Watermarking of Multimedia Contents* 5. Santa Clara, CA: SPIE, 2003.

[16] Kessler, G. C. "An Overview of Steganography for the Computer Forensics Examiner." *Forensic Science Communications* 6 (2004): 1.

[17] Grgic, S., M. Grgic, and M. Mrak. "Reliability of Objective Picture Quality Measures." *J. Electrical Engineering* 55 (2004): 3.

[18] Nill, N. B. "A Visual Model Weighted Cosine Transform for Image Compression and Quality Assessment." *IEEE Trans. Communication* 33 (1985): 551.

[19] Cortes, C., and V. Vapnik. "Support-Vector Networks." *Machine Learning* 20 (1995): 273.

[20] Vapnik, V. *The Nature of Statistical Learning Theory.* New York: Springer-Verlag, 1995.

[21] Burges, C. "A Tutorial on Support Vector Machines for Pattern Recognition." *IEEE Trans. Data Mining and Knowledge Discovery* 2 (1998): 121.

[22] Vapnik, V. *Statistical Learning Theory.* New York: Wiley, 1998.

[23] Heisele, B., et al. "Hierarchical Classification and Feature Reduction for Fast Face Detection with Support Vector Machines." *Pattern Recognition* 36 (2003): 2007.

[24] Shih, F. Y., and S. Cheng. "Improved Feature Reduction in Input and Feature Spaces." *Pattern Recognition* 38 (2005): 651.

[25] Schohn, G., and D. Cohn. "Less Is More: Active Learning with Support Vector Machines." In *Proc. Int. Conf. Machine Learning.* Stanford, CA: Stanford University, 2000.

[26] Syed, N. A., H. Liu and K. K. Sung. "Incremental Learning with Support Vector Machines." In *Proc. Int. Joint Conf. Artificial Intelligence.* Stockholm, 1999.

[27] Campbell, C., N. Cristianini, and A. Smola. "Query Learning with Large Margin Classifiers." In *Proc. Int. Conf. Machine Learning.* Stanford, CA: Stanford University, 2000.

[28] An, J.-L., Z.-O. Wang, and Z.-P. Ma. "An Incremental Learning Algorithm for Support Vector Machine." In *Proc. Int. Conf. Machine Learning and Cybernetics.* Xi'an, China, 2003.

[29] Nguyen, H. T., and A. Smeulders. "Active Learning Using Pre-Clustering." In *Proc. Int. Conf. Machine Learning.* Alberta, 2004.

[30] Mitra, P., C. A. Murthy, and S. K. Pal. "A Probabilistic Active Support Vector Learning Algorithm." *IEEE Trans. Pattern Analysis and Machine Intelligence* 26 (2004): 413.

[31] Cortes, C. and V. Vapnik. "Support-Vector Network." *Machine Learning* 20 (1995): 273.

[32] Platt, J. C., N. Cristianini, and J. Shawe-Taylor. "Large Margin DAGs for Multiclass Classification." In *Advances in Neural Information Processing Systems.* Boston: MIT Press, 2000.

[33] Hsu, C. W., and C. J. Lin. "A Comparison of Methods for Multiclass Support Vector Machines." *IEEE Trans. Neural Networks* 13 (2002): 415.

[34] Herbrich, R. *Learning Kernel Classifiers Theory and Algorithms.* Boston: MIT Press, 2001.

[35] Martinez, A. M., and A. C. Kak. "PCA versus LDA." *IEEE Trans. Pattern Analysis and Machine Intelligence* 23 (2001): 228.

[36] Westfeld, A. "High Capacity Despite Better Steganalysis (F5–A Steganographic Algorithm)." In *Proc. Int. Workshop Information Hiding.* New York, 2001.

[37] Provos, N. "Defending against Statistical Steganalysis." In *Proc. USENIX Security Symposium.* Washington, DC, 2001.

[38] Fridrich, J., M. Goljan, and D. Hogea. "Steganalysis of JPEG Images: Breaking the F5 Algorithm." In *Proc. Int. Workshop Information Hiding.* Noordwijkerhout, Netherlands, 2002, 310.

[39] Alturki, F., and Mersereau, R. "A Novel Approach for Increasing Security and Data Embedding Capacity in Images for Data Hiding Applications." In *Proc. Int. Conf. Information Technology: Coding and Computing.* Las Vegas, NV, 2001.

12 Genetic Algorithm-Based Steganography

A steganographic system must be able to embed hidden messages into a cover-image in such a way that no one can suspect any alteration on the cover-image. Otherwise its goal will be void. Most passive warden distinguishes the stego-images by analyzing their statistic features. Since the steganalytic system analyzes certain statistic features of an image, the idea of developing a robust steganographic system is to generate the stego-image by avoiding changing the statistic features of the cover image. Several studies have presented algorithms for steganographic and steganalytic systems, buy very few have discussed algorithms for breaking steganalytic systems. Recently, Chu et al. [1] have presented a direct cosine transform–based steganographic system by utilizing the similarities of DCT coefficients between the adjacent image blocks where the embedding distortion is spread. Their algorithm can allow random selection of DCT coefficients in order to maintain key statistic features. However, the drawback of their approach is that the capacity of the embedded message is limited—that is, only 2 bits for an 8×8 DCT block.

In this chapter, we present a method for breaking steganalytic systems that is based on genetic algorithm (GA). The emphasis is shifted from traditionally avoiding the change of statistic features to artificially counterfeiting the statistic features. Our idea is based on the following: in order to manipulate the statistic features for breaking the inspection of steganalytic systems, the GA-based approach is adopted to counterfeit several stego-images (candidates) until one of them can break the inspection of steganalytic systems.

This chapter is organized as follows. An overview of a GA-based breaking methodology is given in section 12.1. GA-based breaking algorithms on spatial-domain steganalytic systems (SDSSs) and frequency-domain steganalytic systems (FDSSs) are presented in sections 12.2 and 12.3, respectively. Experimental results are shown in section 12.4, and a complexity analysis is provided in section 12.5.

12.1 AN OVERVIEW OF THE GA-BASED BREAKING METHODOLOGY

The genetic algrithm, introduced in a seminal work by Holland [2], is commonly used as an adaptive approach that provides a randomized, parallel, and global search based on the mechanics of natural selection and genetics in order to find solutions to a problem.

In general, the GA starts with some randomly selected genes as the first generation, called *population*. Each individual in the population corresponding to a solution in the problem domain is called a *chromosome*. An objective, called *fitness function*, is used to evaluate the quality of each chromosome. The chromosomes of high quality will survive and form a new population of the next generation. By using three operators—reproduction, crossover, and mutation—we recombine a new generation

g_0	g_1	g_2	g_3	g_4	g_5	g_6	g_7
g_8	g_9	g_{10}	g_{11}	g_{12}	g_{13}	g_{14}	g_{15}
g_{16}	g_{17}	g_{18}	g_{19}	g_{20}	g_{21}	g_{22}	g_{23}
g_{24}	g_{25}	g_{26}	g_{27}	g_{28}	g_{29}	g_{30}	g_{31}
g_{32}	g_{33}	g_{34}	g_{35}	g_{36}	g_{37}	g_{38}	g_{39}
g_{40}	g_{41}	g_{42}	g_{43}	g_{44}	g_{45}	g_{46}	g_{47}
g_{48}	g_{49}	g_{50}	g_{51}	g_{52}	g_{53}	g_{54}	g_{55}
g_{56}	g_{57}	g_{58}	g_{59}	g_{60}	g_{61}	g_{62}	g_{63}

(a)

0	0	1	0	0	2	0	1
1	2	0	0	1	1	0	1
2	1	1	0	0	-2	0	1
3	1	2	5	0	0	0	0
0	2	0	0	0	1	0	0
0	-1	0	2	1	1	4	2
1	1	1	0	0	0	0	0
0	0	0	0	1	-2	1	3

(b)

FIGURE 12.1 The numbering positions corresponding to 64 genes.

to find the best solution. The process is repeated until a predefined condition is satisfied or a constant number of iterations are reached. The predefined condition in this instance is the situation in which we can correctly extract a hidden message.

In order to apply the GA for embedding messages into the frequency domain of a cover image to obtain the stego-image, we use the chromosome ξ, consisting of n genes as $\xi = g_0, g_1, g_2, \ldots, g_n$. Figure 12.1 gives an example of a chromosome ($\xi \in Z^{64}$) containing 64 genes ($g_i \in Z$ ([integers]). Figure 12.1(a) shows the distribution order of a chromosome in an 8×8 block, and Figure 12.1(b) shows an example of the corresponding chromosome.

The chromosome is used to adjust the pixel values of a cover image to generate a stego-image, so the embedded message can be correctly extracted and at the same time the statistical features left intact in order to break steganalytic systems. A fitness function is used to evaluate the embedded message and statistic features.

Let C and S respectively denote a cover image and a stego-image of size 8×8. We generate the stego-images by adding the cover image and the chromosome as

$$S = \{s_i | s_i = c_i + g_i, \quad \text{where } 0 \le i \le 63\}. \tag{12.1}$$

12.1.1 THE FITNESS FUNCTION

In order to embed messages into DCT-based coefficients and avoid the detection of steganalytic systems, we develop a fitness function to evaluate the following two terms:

1. *Analysis*(ξ, C): The analysis function evaluates the difference between the cover image and the stego-image in order to maintain statistical features. It is related to the type of the steganalytic systems used and will be explained in sections 12.2 and 12.3.
2. *BER*(ξ, C): The bit error rate (BER) sums up the bit differences between the embedded and the extracted messages. It is defined as

$$BER(\xi, C) = \frac{1}{|Message^H|} \sum_{i=0}^{all\ pixels} |Message_i^H - Message_i^H|, \tag{12.2}$$

where $Message^H$ and $Message^E$ denote the embedded and the extracted binary messages, respectively, and $|Message^H|$ denotes the length of the message. For example, if $Message^H = 11111$ and $Message^H = 10101$, then $BER\,(\xi,C) = 0.4$.

We use a linear combination of the analysis and the bit error rate to be the fitness function as

$$Evaluation\,(\xi,C) = \alpha_1 \times Analysis(\xi,C) + \alpha_2 \times BER(\xi,C) \qquad (12.3)$$

where α_1 and α_2 denote weights. The weights can be adjusted according to the user's demand on the degree of distortion to the stego-image or the extracted message.

12.1.2 REPRODUCTION

The formula for *Reproduction* is

$$Reproduction\,(\Psi,k) = \{\xi_i \mid Evaluation(\xi_i,C) \leq \Omega \text{ for } \xi_i \in \Psi\}, \qquad (12.4)$$

where Ω is a threshold for sieving chromosomes, and $\Psi = \{\xi_1, \xi_2, \ldots, \xi_n\}$. It is used to reproduce better k chromosomes from the original population for higher qualities.

12.1.3 CROSSOVER

The formula for crossover is

$$Crossover(\Psi,l) = \{\xi_i \ominus \xi_j \mid \xi_i, \xi_j \in \Psi\} \qquad (12.5)$$

where $\ominus$ denotes the operation to generate chromosomes by exchanging genes from their parents, ξ_i and ξ_j. It is used to gestate l better offsprings by inheriting good genes (i.e., higher qualities in the fitness evaluation) from their parents. The often-used crossovers are *one-point*, *two-point*, and *multipoint*. The criteria for selecting a suitable crossover depend on the length and structure of chromosomes. We adopt the one- and two-point crossovers, as shown in Shih and Wu [3], throughout this chapter.

12.1.4 MUTATION

The formula for mutation is

$$Mutation(\Psi,m) = \{\xi_i \circ j \mid 0 \leq j \leq |\xi_i| \text{ and } \xi_i \in \Psi\} \qquad (12.6)$$

where $\circ$ denotes the operation to randomly select a chromosome ξ_i from Ψ and change the j-th bit from ξ_i. It is used to generate m new chromosomes. The mutation is usually performed with a probability $p\,(0 < p \leq 1)$, meaning only p portion of the genes in a chromosome will be selected to be mutated. Since the length of a chromosome is 64 in this instance, and there are only one or two genes to be mutated when the GA mutation operator is performed, we select p to be 1/64 or 2/64.

Note that in each generation the new population is generated by the above three operations. The new population is actually the same size of $k + l + m$ as the original population. Note also that to break the inspection of the steganalytic systems we use a straightforward GA selection method in which the new generation is generated based on the chromosomes having superior evaluation values in the current generation. For example, only the top 10% of the current chromosomes will be considered for GA operations such as reproduction, crossover, and mutation.

The mutation tends to be more efficient than the crossover if a potential solution is close to the real optimum solution. In order to enhance the performance of our GA-based methodology, we generate a new chromosome with desired minor adjustment when the previous generation is close to the goal. Therefore, the strategy of dynamically determining the ratio of three GA operations is utilized. Let R_P, R_M, and R_C respectively denote the ratios of production, mutation, and crossover. In the beginning, we select a small R_P and R_M and a large R_C to enable the global search. After certain iterations, we will decrease R_C and increase R_M and R_P to shift focuses on a local search if the current generation is better than the old one; otherwise, we will increase R_C and decrease R_M and R_P to enlarge the range to a global search. Note that this property must be satisfied: $R_P + R_M + R_C = 100\%$.

The recombination strategy of our GA-based algorithm is presented below. We apply the same strategy in recombining the chromosome throughout this chapter, except that the fitness function is differently defined with respect to the properties of individual problems.

THE ALGORITHM FOR RECOMBINING THE CHROMOSOMES

1. Initialize the base population of chromosomes R_P, R_M, and R_C.
2. Generate candidates by adjusting the pixel values of the original image.
3. Determine the fitness value of each chromosome.
4. If a predefined condition is satisfied or a constant number of iterations are reached, the algorithm will stop and output the best chromosome to be the solution; otherwise, go to the following steps to recombine the chromosomes.
5. If a certain number of iterations are reached, go to step 6 to adjust R_P, R_M, and R_C; otherwise, go to step 7 to recombine the new chromosomes.
6. If 20% of the new generation are better than the best chromosome in the preceding generation, then $R_C = R_C - 10\%$, $R_M = R_M + 5\%$ and $R_P = R_P + 5\%$; otherwise, $R_C = R_C + 10\%$, $R_M = R_M - 5\%$ and $R_P = R_P - 5\%$.
7. Obtain the new generation by recombining the preceding chromosomes using production, mutation, and crossover.
8. Repeat step 2.

12.2 THE GA-BASED BREAKING ALGORITHM ON THE SDSS

As has been mentioned, in order to generate a stego-image to pass though the inspection of the SDSS, the messages should be embedded into the specific positions for maintaining the statistical features of a cover image. In the spatial-domain embedding approach, it is difficult to select such positions since the messages are distributed regularly. On the other hand, if the messages are embedded into specific positions of coefficients of a transformed image, the changes in the spatial domain are difficult to predict. This section intends to find the desired positions on the frequency domain that produce minimum statistic features disturbance on the spatial domain. The GA is applied in order to to generate the stego-image by adjusting the pixel values on the spatial domain using the following two criteria:

1. Evaluate the extracted messages obtained from the specific coefficients of a stego-image to ascertain that they are as close as possible to the embedded messages.
2. Evaluate the statistical features of the stego-image and compare them with those of the cover image such that the differences should be as few as possible.

12.2.1 GENERATING THE STEGO-IMAGE ON THE VISUAL STEGANALYTIC SYSTEM

Westfeld and Pfitzmann have presented a visual steganalytic system (VSS) that uses an assignment function of color replacement, called the *visual filter*, to efficiently detect a stego-image by translating a grayscale image into a binary one [4]. Therefore, in order to break the VSS, the two results, VF^C and VF^S, of applying the visual filter on the cover image and stego-image, respectively, should be as identical as possible. The VSS was originally designed to detect GIF-format images by reassigning the color in the color palette of an image. We extend this method to detect the BMP format images as well by setting the odd- and even-numbered grayscales to black and white, respectively. The $Analysis(\xi, C)$ to the VSS indicating the sum of difference between LSB^C and LSB^S is defined as

$$Analysis(\xi, C) = \frac{1}{|C|} \sum_{i=0}^{all\ pixels} \left(VF_i^C \oplus VF_i^S \right) \tag{12.7}$$

where $\oplus$ denotes the Exclusive-OR (XOR) operator. Our algorithm is delineated below.

THE ALGORITHM FOR GENERATING A STEGO-IMAGE ON THE VSS

1. Divide a cover image into a set of cover images of size 8×8.
2. For each 8×8 cover image, we generate a stego-image based on the GA to perform the embedding procedure, as well as to ensure that LSB^C and LSB^S are as identical as possible.
3. Combine all of the 8×8 stego-images together to form one complete stego-image.

12.2.2 GENERATING THE STEGO-IMAGE ON THE IMAGE QUALITY MEASURE-BASED STEGANALYTIC SYSTEM

Avcibas and colleagues have proposed a steganalytic system by analyzing the image quality measures (IQMs) of the cover image and the stego-image [5–6]. The IQM-based spatial-domain steganalytic system (IQM-SDSS) consists of two phases: training and testing. In the training phase, the IQM is calculated between an image and its filtered image using a low-pass filter based on the Gaussian kernel. If there are N images and q IQMs in the training set, let x_{ij} denote the score in the i-th image and the j-th IQM, where $1 \leq i \leq N$, $1 \leq j \leq q$. Let y_i be the value of -1 or 1, indicating the cover image or stego-image, respectively. We can represent all the images in the training set as

$$
\begin{aligned}
y_1 &= \beta_1 x_{11} + \beta_2 x_{12} + \cdots + \beta_q x_{1q} + \varepsilon_1 \\
y_2 &= \beta_1 x_{21} + \beta_2 x_{22} + \cdots + \beta_q x_{2q} + \varepsilon_2 \\
&\;\;\vdots \\
y_N &= \beta_1 x_{N1} + \beta_2 x_{N2} + \cdots + \beta_q x_{Nq} + \varepsilon_N
\end{aligned}
\tag{12.8}
$$

where $\varepsilon_1, \varepsilon_2, \ldots, \varepsilon_N$ denote random errors in the linear regression model [7]. The linear predictor $\beta = [\beta_1, \beta_2, \ldots, \beta_q]$ can then be obtained from all of the training images.

In the testing phase, we use the q IQMs to compute y_i to determine whether it is a stego-image. If y_i is positive, the test image is a stego-image; otherwise, it is a cover image.

We first train the IQM-SDSS in order to obtain the linear predictor—that is, $\beta = [\beta_1, \beta_2, \ldots, \beta_q]$—from our database. Then, we use β to generate the stego-image via our GA-based algorithm, so that it can pass through the inspection of IQM-SDSS. Note that the GA procedure is not used in the training phase. The $Analysis(\xi, C)$ to the IQM-SDSS is defined as

$$
Analysis(\xi, C) = \beta_1 x_1 + \beta_2 x_2 + \cdots + \beta_q x_q
\tag{12.9}
$$

THE ALGORITHM FOR GENERATING A STEGO-IMAGE ON THE IQM-SDSS

1. Divide a cover image into a set of cover images of size 8×8.
2. Adjust the pixel values in each 8×8 cover image based on GA to embed messages into the frequency domain and ensure that the stego-image can pass through the inspection of the IQM-SDSS. (The procedures for generating the 8×8 stego-image are presented below.)
3. Combine all of the 8×8 embedded images together to form one completely embedded image.
4. Test the embedded image on the IQM-SDSS. If it passes, it is the desired stego-image; otherwise, repeat steps 2–4 until a desired stego-image is found.

THE PROCEDURE FOR GENERATING AN 8×8 EMBEDDED IMAGE ON THE IQM-SDSS

1. Define the fitness function, the number of genes, the size of population, the crossover rate, the critical value, and the mutation rate.
2. Generate the first generation by a random selection.

3. Generate an 8×8 embedded-image based on each chromosome.
4. Evaluate the fitness value for each chromosome by analyzing the 8×8 embedded-image.
5. Obtain the better chromosome based on the fitness value.
6. Recombine new chromosomes by crossover.
7. Recombine new chromosomes by mutation.
8. Repeat steps 3–8 until a predefined condition is satisfied or a constant number of iterations are reached.

12.3 THE GA-BASED BREAKING ALGORITHM ON THE FDSS

Fridrich, Goljan, and Hogea have presented a steganalytic system for detecting JPEG stego-images based on the assumption that the histogram distributions of some specific alternating current (AC) DCT coefficients of a cover image and its cropped image should be similar [8]. Note that in the DCT coefficients, only the zero frequency (0,0) is the direct current (DC) component, and the remaining frequencies are the AC components. Let $h_{kl}(d)$ and $\bar{h}_{kl}(d)$ respectively denote the total number of AC DCT coefficients in the 8×8 cover image and its corresponding 8×8 cropped image with the absolute value equal to d at location (k, l), where $0 \leq k, l \leq 7$. Note that the 8×8 cropped image, defined in Fridrich et al. [8], are obtained using the same way as the 8×8 cover image, with a horizontal shift by 4 pixels.

The probability, ρ_{kl}, of the modification of a nonzero AC coefficient at (k, l) can be obtained by

$$\rho_{kl} = \frac{\bar{h}_{kl}(1)[h_{kl}(0) - \bar{h}_{kl}(0)] + [h_{kl}(1) - \bar{h}_{kl}(1)][\bar{h}_{kl}(2) - \bar{h}_{kl}(1)]}{[\bar{h}_{kl}(1)]^2 + [\bar{h}_{kl}(2) - \bar{h}_{kl}(1)]^2} \quad (12.10)$$

Note that the final value of the parameter ρ is calculated as an average over the selected low-frequency DCT coefficients $(k, l) \in \{(1,2), (2,1), (2,2)\}$, and only 0, 1, and 2 are considered when checking the coefficient values of the specific frequencies between a cover and its corresponding cropped image.

Our GA-based breaking algorithm on the JPEG frequency-domain steganalytic system (JFDSS) is intended to minimize the differences between the two histograms of a stego-image and its cropped image. It is presented below.

THE ALGORITHM FOR GENERATING A STEGO-IMAGE ON THE JFDSS

1. Compress a cover image via JPEG and divide it into a set of small cover images of size 8×8. Each is transformed via DCT.
2. Embed the messages into the specific DCT coefficients and decompress the embedded image by inverse DCT.
3. Select a 12×8 working window and generate an 8×8 cropped image for each 8×8 embedded image.
4. Determine the overlapping area between each 8×8 embedded image and its cropped image.

5. Adjust the overlapping pixel values by making sure that the coefficients of some specific frequencies (k, l) of the stego-image and its cropped image are as identical as possible and the embedded messages are not altered.
6. Repeat steps 3–5 until all the 8×8 embedded images are generated.

Let $Coef^{Stego}$ and $Coef^{Crop}$ denote the coefficients of each 8×8 stego-image and its cropped image. The $Analysis(\xi, C)$ to the JFDSS is defined as

$$Analysis(\xi, C) = \frac{1}{|C|} \sum_{i=0}^{all\ pixels} \left(Coef_i^{Stego} \otimes Coef_i^{Crop} \right), \qquad (12.11)$$

where $\otimes$ denotes the operator defined by

$$\begin{cases} Coef_i^{Stego} \otimes Coef_i^{Crop} = 1 & \text{if } (Coef_i^{Stego} = 0 \text{ and } Coef_i^{Crop} \neq 0) \text{ or} \\ & \quad (Coef_i^{Stego} = 1 \text{ and } Coef_i^{Crop} \neq 1) \text{ or} \\ & \quad (Coef_i^{Stego} = 2 \text{ and } Coef_i^{Crop} \neq 2) \text{ or} \\ & \quad (Coef_i^{Stego} \neq 0, 1, 2 \text{ and } Coef_i^{Crop} = 0, 1, 2) \\ Coef_i^{Stego} \otimes Coef_i^{Crop} = 0 & \text{otherwise} \end{cases} \qquad (12.12)$$

Note that, 0, 1, and 2 denote the values in the specific frequencies obtained by dividing the quantization table. We only consider the values of the desired frequencies to be 0, 1, 2, or some values in equation 12.12 because of the strategy of the JFDSS in equation 12.10.

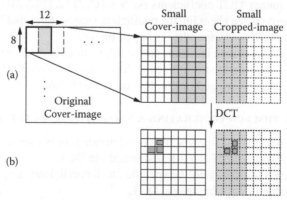

The Frequency Domain of Cover Image and Cropped Images

FIGURE 12.2 An example of our GA-based algorithm on the JFDSS.

Figure 12.2 shows an example of our GA-based algorithm on the JFDSS. In Figure 12.2(a), we select a 12×8 working window for each 8×8 stego-image, and generate its 8×8 cropped image. Note that the shaded pixels indicate their overlapping area, and the three black boxes in Figure 12.2(b) are the desired locations.

12.4 EXPERIMENTAL RESULTS

This section provides experimental results to show that our GA-based steganographic system can successfully break the inspection of steganalytic systems. For testing our algorithm, we use a database of 200 grayscale images of size 256×256. All the images were originally stored in the BMP format.

12.4.1 THE GA-BASED BREAKING ALGORITHM ON THE VSS

We test our algorithm on the VSS. Figures 12.3(a) and 12.3(f) show a stego-image and a message image of sizes 256×256 and 64×64, respectively. We embed 4 bits into the 8×8 DCT coefficients on frequencies (0,2), (1,1), (2,0), and (3,0) to avoid distortion. Note that the stego-image in Figure 12.3(a) is generated by embedding Figure 12.3(f) into the DCT coefficients of the cover image using our GA-based algorithm. Figures 12.3(b)–12.3(j) display the bit planes from 7 to 0. Figure 12.4 shows the stego-image and its visual-filtered result. It is difficult to determine that Figure 12.4(a) is a stego-image.

Figure 12.5 shows the relationship of the average iteration for adjusting an 8×8 cover image versus the correct rate of the visual filter. The correct rate is the percentage of similarity between the transformed results of the cover image and the stego-image using the visual filter. Note that the BERs in Figure 12.5 are all 0%.

12.4.2 THE GA-BASED BREAKING ALGORITHM ON THE IQM-SDSS

We generate three stego-images as the training samples for each cover image with the Photoshop plug-in Digimarc [9], Cox's technique [10], and S-Tools [11]. Therefore, there are a total of 800 images of size 256×256, including 200 cover images and 600 stego-images. The embedded message sizes are 1/10, 1/24, and 1/40 of the cover image size for Digimarc, Cox's technique, and S-tools, respectively. Note that the IQM-SDSS can detect the stego-images containing the message size of 1/100 of the cover image [5]. We develop the following four training strategies to obtain the linear predictors:

1. Train all the images in the database to obtain the linear predictor β^A.
2. Train 100 cover images and 100 stego-images to obtain the linear predictor β^B, in which the stego-images are obtained via Cox's technique.
3. Train 100 cover images and 100 stego-images to obtain the linear predictor β^C, in which the stego-images are obtained by the Photoshop plug-in Digimarc.
4. Train 100 cover images and 100 stego-images to obtain the linear predictor β^D, in which the stego-images are obtained by S-Tools.

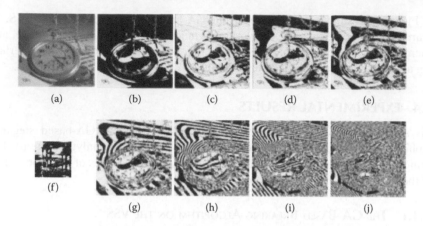

FIGURE 12.3 A stego-image generated by our GA-based algorithm and its 8-bit planes.

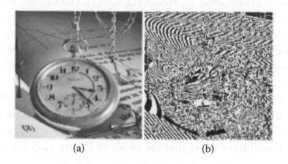

FIGURE 12.4 A stego-image and its visual-filtered result.

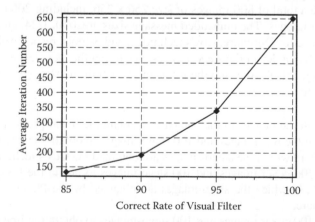

FIGURE 12.5 The relationship of the average iteration versus the correct rate of the visual filter.

TABLE 12.1
The Experimental Results of the GA-Based Algorithm on the IQM-SDSS

		β^A	β^B	β^C	β^D	Average
50 SI_A images	FN		94 %	94 %	100 %	96 %
50 SI_B images	FN	84 %		84%	80 %	83 %
50 SI_C images	FN	82 %	86 %		88 %	85 %
50 SI_D images	FN	82 %	86 %	82 %		83 %
50 cover images	FP	6 %	10 %	16 %	16 %	12 %

In the testing phase, we generate 50 stego-images for each linear predictor. Therefore, we will obtain four types of stego-images, SI_A, SI_B, SI_C, and SI_D, corresponding to β^A, β^B, β^C, and β^D, respectively. It is obvious that all of the SI_A images will pass through the inspection of the IQM-SDSS with the linear predictor β^A, but may fail with other linear predictors; the same is true for the SI_B, SI_C, and SI_D as well.

Table 12.1 shows the false negative (Type I error) and false positive (Type II error) rates of testing each type of stego-images and cover images under different linear predictors. For example, the false negative (FN) rate of testing 50 SI_A stego-images under β^B is 94%, indicating that the β^B-IQMSS falsely decides that 94% of our stego-images belong to the cover image. On the other hand, the false positive (FP) rate of testing 50 cover images under β^B is 10%, indicating that the β^B-IQMSS decides that 10% of our cover images belong to the stego-images. Note that the cover images used in the testing phase are different from those used in the training phase.

12.4.3 THE GA-BASED BREAKING ALGORITHM ON THE JFDSS

Figure 12.6 shows an example of adjusting an 8×8 embedded image to obtain an 8×8 stego-image for breaking the JFDSS. Figure 12.6(a) shows a 12×8 working window where the enclosed is the overlapping area. Figures 12.6(b) and 12.6(c) show the original 8×8 embedded images and cropped images, respectively. We embed 1 on (1,2), (2,1), and (2,2) by compressing Figure 12.6(b) using JPEG under 70% compression quality to obtain Figure 12.6(d). Note that the top left pixel is (0,0) and the messages can be embedded into any frequency of the transformed domain, so that the embedding capacity could be sufficiently high. Due to the fact that the JFDSS only checks frequencies (1,2), (2,1), and (2,2), we show an example of embedding 3 bits into these three frequencies. Similarly, Figure 12.6(e) is obtained by compressing Figure 12.6(c) using JPEG under the same compression quality. By evaluating the frequencies (1,2), (2,1), and (2,2), the JFDSS can determine whether the embedded

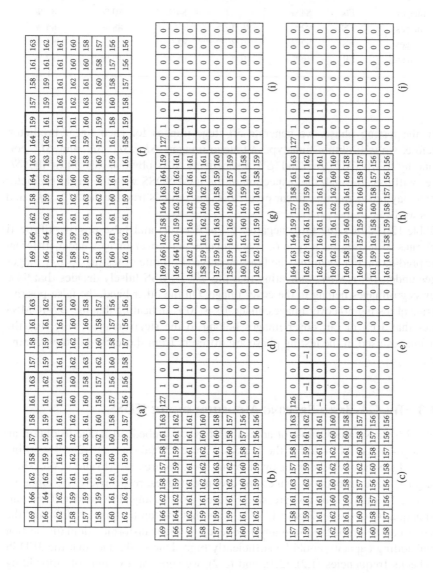

FIGURE 12.6 An example of adjusting an 8 × 8 embedded image.

16	11	10	16	24	40	51	61
12	12	14	19	26	58	60	55
14	13	16	24	40	57	69	56
14	17	22	29	51	87	80	62
18	22	37	56	68	109	103	77
24	35	55	64	81	104	113	92
49	64	78	87	103	121	120	101
72	92	95	98	112	100	103	99

(a)

10.1	7.1	65	10.1	14.9	24.5	31.1	37.1
7.7	7.7	89	11.9	16.1	35.3	36.5	33.5
8.9	83	10.1	14.9	24.5	34.7	41.9	34.1
8.9	10.7	13.7	17.9	31.1	52.7	48.5	37.7
11.3	13.7	22.7	34.1	41.3	65.9	62.3	46.7
14.9	21.5	33.5	38.9	49.1	62.9	68.3	55.7
29.9	38.9	47.3	52.7	62.3	73.1	72.5	61.1
43.7	55.7	57.5	59.3	67.7	60.5	62.3	59.9

(b)

FIGURE 12.7 The quantization table of JPEG compression.

image is a stego-image. Therefore, in order to break the JFDSS, we obtain the stego-image shown in Figure 12.6(f). Figures 12.6(g) and 12.6(h) show the new embedded images and cropped images, respectively. Similarly, we obtain Figures 12.6(i) and 12.6(j) by compressing Figures 12.6(g) and 12.6(h), respectively, using JPEG under 70% compression quality. Therefore, the JFDSS cannot distinguish from the frequencies (1,2), (2,1), and (2,2).

Let $QTable(i, j)$ denote the standard quantization table, where $0 \le i, j \le 7$. The new quantization table, $NewTable(i, j)$, with $x\%$ compression quality, can be obtained by

$$NewTable(i, j) = \frac{QTable(i, j) \times factor + 50}{100} \qquad (12.13)$$

where the factor is determined by

$$\begin{cases} factor = \dfrac{5000}{x} & \text{if } x \le 50 \\ factor = 200 - 2x & \text{otherwise} \end{cases} \qquad (12.14)$$

Figures 12.7(a) and 12.7(b) show the quantization tables of the standard and 70% compression quality, respectively.

12.5 COMPLEXITY ANALYSIS

In general, the complexity of our GA-based algorithm is related to the size of the embedded message and the position of embedding [12]. Figure 12.8 shows the relationship between the embedded message and the required iterations in which the

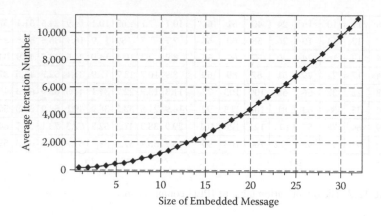

FIGURE 12.8 The relationship between the size of the embedded message and the required iterations.

cover image is of size 8 × 8 and the message is embedded into the least significant bit of the cover image in a zigzag order starting from the DC (zero-frequency) component. We observe that the more messages that are embedded in a stego-image, the more iterations that are required in our GA-based algorithm. Figure 12.9 shows an example in which we embed a 4-bit message into a different bit plane of DCT coefficients. We observe that the lower the bit plane that is used for embedding, the more iterations that are required in our GA-based algorithm.

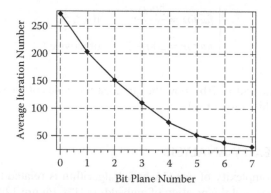

FIGURE 12.9 Embedding a 4-bit message into a different bit plane.

REFERENCES

[1] Chu, R., et al. "A DCT-Based Image Steganographic Method Resisting Statistical Attacks." In *Intl. Conf. Acoustics, Speech, and Signal Processing.* Montreal, Quebec, 2004.

[2] Holland, J. H. *Adaptation in Natural and Artificial Systems.* Ann Arbor: University of Michigan Press, 1975.

[3] Shih, F. Y. and Y.-T. Wu. "Enhancement of Image Watermark Retrieval Based on Genetic Algorithm." *J. Visual Commu. and Image Repr.,* 16 (2005): 115.

[4] Westfeld, A., and A. Pfitzmann. "Attacks on Steganographic Systems Breaking the Steganographic Utilities EzStego, Jsteg, Steganos, and S-Tools and Some Lessons Learned." In *Proc. Int. Workshop Information Hiding.* Dresden, Germany, 1999, 61.

[5] Avcibas, I., N. Memon, and B. Sankur. "Steganalysis Using Image Quality Metrics." *IEEE Trans. Image Processing* 12 (2003): 221.

[6] Avcibas, I., and Sankur, B. "Statistical Analysis of Image Quality Measures." *J. Electron. Imag.* 11 (2002): 206.

[7] Rencher, A. C. *Methods of Multivariate Analysis.* New York: Wiley, 1995.

[8] Fridrich, J., M. Goljan, and D. Hogea. "New Methodology for Breaking Steganographic Techniques for JPEGs." In *Proc. SPIE,* Santa Clara, CA: SPIE, 2003.

[9] PictureMarc, Embed Watermark, v 1.00.45, Digimarc Corporation.

[10] Cox, I., et al. "Secure Spread Spectrum Watermarking for Multimedia." *IEEE Trans. Image Processing* 6 (1997): 1673.

[11] Brown, A. *S-Tools for Windows.* Shareware application, 1994. Available online at ftp://idea.sec.dsi.unimi.it/pub/security/crypt/code/s-tools4.zip.

[12] Shih, F. Y., and S. Y. Wu. "Combinational Image Watermarking in the Spatial and Frequency Domains." *Pattern Recognition* 36 (2003): 969.

Index

T - #0191 - 101024 - C0 - 229/152/11 [13] - CB - 9781420047578 - Gloss Lamination